Monkeys Don't Wear Diapers

Monkeys Don't Wear Diapers

Heartwarming and Heartbreaking Stories
from a Monkey Sanctuary

POLLY SCHULTZ

WITH KENNETH LITWAK

Animal Welfare Institute

The Animal Welfare Institute is a non-profit charitable organization founded in the United States in 1951 and dedicated to reducing animal suffering caused by people. AWI engages policymakers, scientists, industry, and the public to achieve better treatment of animals everywhere—in the laboratory, on the farm, in commerce, at home, and in the wild.

Published by the Animal Welfare Institute
900 Pennsylvania Avenue, SE, Washington, DC 20003
www.awionline.org

Monkeys Don't Wear Diapers: Heartwarming and Heartbreaking Stories from a Monkey Sanctuary / by Polly Shultz; with Kenneth Litwak

ISBN 978-0-938414-80-3
LCN 2015930227

Printed in the United States
Edited by Dave Tilford and Cathy Liss
Design by Ava Rinehart and Alexandra Alberg
Cover photo: Ernie

I dedicate this book to the beautiful souls:
monkeys who call OPR home, those yet to be rescued, and the
compassionate humans whose continued love and support
make OPR a successful, safe and happy place.

Contents

Foreword

This book is about a special group of monkeys. Some came from misinformed owners, who tried to give them a good home, but were overwhelmed by the needs of a monkey. Others were horribly mistreated by abusive owners, who faced criminal charges for their actions. Still others were retired from research institutions, after they were no longer needed for experiments.

The legal and illegal trade in monkeys is staggering. An Internet search using the phrase, "baby monkey for sale," reveals thousands of ads for virtually every known species of monkey. For a few hundred dollars, a person can purchase a monkey.

It is a terrible decision.

These monkeys are almost always condemned to a horrible life; kept in completely unsuitable, often unimaginably grim conditions.

I have learned a tremendous amount from monkeys over nearly 20 years of providing sanctuary for them. Each monkey is an individual who has specific needs, desires, and aspirations; not so different from people. They require a tremendous amount of work, love, patience, and attention, which I willingly give them. In return, they accept me into their world and I come out a much richer person.

I don't know how I can find words strong enough to thank those who have helped make this dream a reality, and who have helped make OPR Coastal Primate Sanctuary (OPR) a safe and happy place for so many monkeys. In our growing and hurried world, finding compassion for other people can often be a difficult task. Providing the same compassion to animals can be even more difficult. Yet, when it would be easy to look the other way, so many people have reached out, helped us, and continue to make a difference.

You are angels and will always be heroes to the animals and the people who care for them.

There are so many of you who have made a difference. It would take an entire book were I to thank all of you, individually. Know that your contributions have had a tremendous impact on our ability to continue providing sanctuary to these needy monkeys.

A special thank you to all of our volunteers who have spent countless hours at OPR maintaining, cleaning, and building habitats... and just caring.

And thank you:

Cathy Liss, Dave Tilford, Ava Rinehart, Alexandra Alberg, and the staff of the Animal Welfare Institute for making the creation of this book possible. Kenneth Litwak, my co-author, for your compassion, expertise, and dedication. Viktor Reinhardt and all of the LAREF members for many years of encouragement, valuable advice, and support. Drs. Fay Rankin, Robert MacArthur, Mark Stoenner, Margaret Wissman, Robin Lane, and all of your veterinary staff, for the years of expertise, support, and advice, as you cared for all of the monkeys at OPR. Don, Arlene and Jonah Lee for believing in me and for graciously donating the breathtaking land and buildings for our sanctuary, and for your continued love and support. George and Evelyn Chalustowski for your love, compassion and years of support. Damon, Daylee and Kiersten Shaw, Kelly Gutierrez, and the Douglas and Gloria Rumberger Foundation for caring and for helping us make such a positive difference. Robert Reichers and Jann Dryer, whose generous act of kindness to the OPR mission will be honored and remembered through the success of every primate's life in which we are able to alleviate suffering. Jean Barr, Jody Gurin, Love INC, Home Depot, Unchained Brotherhood, and so many more! All are heroes to OPR.

And a special thank you to everyone who loves monkeys, has thought of owning one, and decides, after reading this book, to let them remain wild and free.

Polly Schultz

"From a very early age, I have always had a
deep connection with animals. I have watched
them, interacted with them, understood them,
and wanted to protect them from harm."

Polly

ANIMALS have always been part of my life. There has hardly ever been a time when I didn't feel a deep love and connection with them. My fondest memories as a child are of times when I helped animals or shared some experience with them.

My parents would laugh as they retold stories of my escapades with animals. When I was barely 3 years old, after playing in the backyard all morning, I came into the house clutching two fuzzy caterpillars in my tiny hands. I put them on the living room floor and watched intently as they crawled around. After naming them and playing with them for over an hour, I decided they needed to have a place to sleep. At this point, my mother stepped in and gently told me that my new "pets" had families who probably missed them. It would be best to return them to the yard, so they could rejoin those families. This set off another round of conversations between me and the caterpillars, as only a 3-year-old can have. I gently picked them up, one in each hand, and headed for the back door, only to be stymied when I realized I no longer had a hand available to open the door. Not to be dissuaded, I carefully put one of the caterpillars in my mouth, freeing my hand to open the door and allowing me to get the caterpillars back to their families.

One of my earliest memories is from a visit to the zoo with my parents. When we reached the monkeys, I stood fascinated, watching them with rapt attention, seemingly for hours. The baby monkeys really captivated me, as they clung so tightly to their mothers' bellies. The affection lavished by the mothers upon their infants, as they nursed and cuddled them, was so tender and loving. I remember thinking that the monkeys were treating their babies the same way I had seen human mothers treat their babies.

We lived only a couple of blocks from a veterinary clinic. When I was 11, I would go there every day after school to make myself available, just in case they needed my "help." After months of offering this help, I must have finally worn them down and they started letting me clean cages and pick up doughnuts in the morning from the store next door. I was so proud of my job. Even if I wasn't making any money, I got to work with animals.

My "job" at the clinic soon came in very handy, after I found three baby birds who had fallen from a nest in the nearby strawberry field. I had put them in a small box, lined with fresh grass, but didn't quite know what to do with them. The veterinarian was so kind as he explained what I would need to do to care for them, how often they would need to be fed (very often), and what food they needed. For the next two weeks, my fingernails were perpetually covered in dirt, from digging up worms and beetles. Yet, my little brood grew quickly and a couple of weeks later I started giving them "flying lessons," running around the yard flapping my arms like crazy. I felt such a tremendous sense of accomplishment. In my mind, I was actually teaching them to fly. Then, when they finally flew away, my heart swelled with pride, as I relived every moment of our short time together; the first of many rescues and rehabilitations.

When I married and was blessed with my children, Michelle and Dustin, I took a brief break from the paid workforce to devote my time to parenting. I loved my children so much and I wanted them to be able to share the joy that animals had always brought to my life. I raised my children to show care and respect for all animals, to understand that animals had feelings and that they were to be treated with kindness. This meant that animals were always to be a large part of their lives as they grew up.

One day, with my children in tow, we went to a city park to feed and watch the ducks. When we arrived, we noticed a mallard duck resting near a picnic table. We walked closer, but he just stayed there, making no effort to get away; not normal behavior. When we got even closer, we quickly realized the reason. One leg was very swollen. From the position of the leg, I knew it was broken. I couldn't leave him there; so our visit to the park ended before it began, as we wrapped him in a jacket and drove him over to a veterinary clinic. I was thrilled when I found out that the leg could be fixed, but then realized that I was going to have to care for "Daffy" while he had a cast on his leg and until he could safely walk and his leg had fully healed. For the next two months, until his release, Daffy lived in our back

LEFT *When my children and I discovered a mallard duck at the park with a broken leg, we took him to a veterinarian. "Daffy" spent six weeks in a cast, waddling around our house and back yard. After two months with us, he was released back into the wild.*

RIGHT *Goober the crow was brought to me as a tiny chick and required constant care and feeding. After he was released, he would still come to greet me every morning, landing on my arm and hoping for a mealworm treat.*

yard, quacking furiously whenever he was hungry and, after the cast came off, swimming joyfully in the little wading pool we set up for him. Our accidental backyard wildlife rehabilitation center had officially opened. Daffy was the first. Word spread quickly of my inability to say "no" to animals in need. Soon, there was a steady stream of injured raccoons and squirrels, orphaned fawns, and more baby birds than I could count. Thankfully, the veterinarian who repaired Daffy's leg continued to help me and soon became one of my dearest friends. His generosity with his time and efforts knew no bounds. We often spoke of the many wild animals he had worked with in the past, including my favorites, monkeys. It was one of those conversations when he told me the story of a Barbary macaque named Amy, who he had saved after she had drunk from a bottle of Drano. Little did I know at the time that this same macaque would end up at my sanctuary 20 years later!

With my children now in school, I started working as a veterinary technician, a job that provided both income and a source of information and supplies for my ever-growing wildlife rehabilitation center. Wildlife rehabilitation had become my real passion and I decided to take advanced wildlife rehabilitation courses, offered through the Oregon Institute of Technology. The courses were difficult, but I devoured the information they gave me and was constantly going to the library to find more about each species I was likely to see. I learned how to properly care for the animals, so they would have the best opportunity to be released. I knew I had found my calling. After successfully completing the courses, I passed the exams given by the US Fish and Wildlife Service and was presented with a state license to care for wild birds, and a federal license to care for fur-bearing mammals.

With every rescue, I gained more experience and was able to successfully rehabilitate and release a greater variety of wildlife. I gave each animal as much care and support as they needed. In return, some would provide me joy long after their release. Goober the crow, who fell from his nest, injuring his wing, would never stray far from me, even after he was released. For years, he would fly over from blocks away and begin cawing to get my attention for a treat. Others were happy to be healed and released, such as the coyote I rescued from a steel-jaw trap and nursed through a broken leg. Any time I went near him, he tried to bite me. Fortunately, I was never bitten and six weeks after arriving, I released him. A quick look back at me and then he vanished into the undergrowth. Soon, I had one of the highest release rates in the state of Oregon and was asked by the local humane society to teach a class for others wanting to become involved in wildlife rehabilitation, gaining many new recruits to help with the animals in my care.

Over the course of 13 years, my wildlife center grew into a successful facility, with a strong board of directors and a flotilla of well-trained rehabilitators. It was at this point that I made the difficult decision to step away from overseeing the wildlife center. I was confident that it was in good hands and I was always available to help, but I thought it was time for me to pursue a career with a steady income. This led me to pharmacy school and a brief career as a pharmacy technician. The income and benefits were good, yet my heart longed to work with the animals. During my breaks at the pharmacy, I would scan the newspapers, looking at the pets for sale. I didn't want to buy a pet, but from my experiences with the humane society I knew these ads were often a cry for help. People rarely sold their pets to make money. They sold their pets because they could no longer care for them and provide them with a good home. I felt like they were calling for me to help them.

And so it was one day, seven years after leaving my wildlife center, *I read the newspaper...*

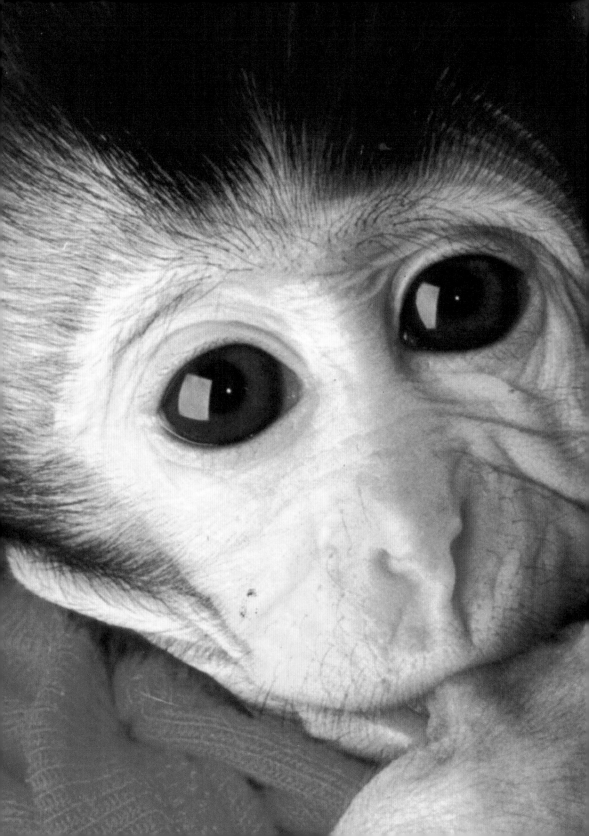

Ernie

I COULDN'T believe it. I must have read and re-read the newspaper ad 20 times and I still couldn't believe it. Why would anyone be offering a baby monkey for sale? There had to be a mistake. Monkeys aren't pets. They're wild animals.

I was shocked and concerned, yet intrigued, as I thought to myself, "What a bad idea. What a very, very bad idea." A thousand reasons for turning the page raced through my head, but I knew that if I didn't answer the ad someone else was going to take that baby monkey. Someone who wouldn't care for him the way I knew I would.

I didn't have children living at home, so I could devote my full attentions to him. I owned my home in the country, so I had the space. I was experienced in the care and rehabilitation of wild animals and had raised nearly every kind of animal. How different could it be to raise a baby monkey? I had nurtured a fascination for monkeys since I was a little child. I just knew I could do it. I was already imagining how I would care for this helpless infant, what I would feed him, where I would keep him, and how I would play with him.

It was too simple. Within minutes of reading the ad, I had picked up the telephone and called the phone number. By later that day, this "very bad idea" was being shipped to me and I was frantically trying to figure out how to raise a baby monkey.

I often look back to that moment and wonder if I had known then what I know now, would I have still made the telephone call? Even if I knew that he would weigh only 8 ounces when he arrived at the house? Even if I knew that he would be so young that his umbilical cord was still attached? Even if I knew that he had a neurological disorder that would result in life-long seizures? Even if I knew that as a result of thoughtless inbreeding, his heart was 4½ times larger than normal, his lungs were underdeveloped, and he had liver problems? Even if I knew that he would be the first of so many monkey rescues that would keep me so busy I wouldn't be able to take a vacation for the next two decades?

The answer is always "Yes."

Then he arrived. When I picked him up at the airport, I couldn't help wondering what kind of person would ship such a tiny, helpless baby monkey in a container, all by himself. He was practically newborn. It was remarkable that he had even survived the plane flight. The person who had arranged for his delivery told me that he came from a dealer who was inbreeding monkeys to create designer, "pocket-sized" monkeys. Her only advice was to treat the baby monkey the same as a human infant—in retrospect, both the best and worst possible advice to raise a baby monkey.

I looked at this fragile, infant monkey and he looked at me with the biggest and most soulful eyes. My heart melted. Quickly, I gave him the warm bottle of formula, which I had brought to the airport, and burped him, as I had done for my own children. As we drove home, he snuggled into my arms and drifted off to sleep. As I looked down at him, I watched his little eyes dance under his closed eyelids. He was dreaming.

What was he dreaming about? Was he dreaming of his mother? She should have been holding him, nursing him, and comforting him. Did she fight and scream when he was taken from her? Was she grieving for the loss of her baby? As I held him and thought of these many questions, I felt overwhelming guilt for taking him from his mother, even as I knew in my heart that he had been taken from his mother before I ever called about him. At that moment, I resolved to make sure he was loved, well cared for, and happy for the rest of his life. I wondered about the other baby monkeys taken from their mothers, and knew that I had to help them, too. At that time, I had no idea of the scale and effects of the trade in monkeys as pets. I just knew I had to do something.

His tiny hand—so incredibly human-like—found its way into his mouth and he was soon sucking his thumb as he slept. I couldn't stop looking at his hands or tracing my finger around his amazing little fingernails. He was so similar to a baby human. He had almost no body hair, except for a thick tuft of shiny, black hair on the top of his head. It was no wonder that people were so enthralled with baby monkeys that they were willing to take them from their mothers, in their misguided attempts to turn them into household pets.

From the moment he arrived, little Ernie had constant health problems, none of which I knew about when he first came to us. When he was only a month old, Ernie started having seizures. It was so frightening to watch. We were constant visitors to our veterinarian. It was during those visits that, in addition to his neurologic problems, we found that Ernie had an enlarged heart and a compromised liver. Even facing such long odds, we knew we would continue to do everything possible to give him a good life, until we couldn't.

By the time Ernie was 3 months old, he was having up to 30 grand mal seizures daily, as well as many petit mal seizures in between. We had been trying multiple medications, in hopes of finding one that worked. It was with some reluctance that our veterinarian gave us phenobarbital; knowing that it would probably stop his seizures, but would also eventually damage his liver even more. With nothing left to try, we started the drug and the results were astounding. Instead of near-constant seizure activity, the frequency decreased to twice weekly, then once monthly, eventually settling to once every few months as he got older. It was the lesser of two evils, but at least we were able to give him a better quality of life now.

By the time he was 6 months old, Ernie had already outlived the predictions of his health-care team of veterinarians, neurologists, and cardiologists. It was a grueling and exhausting time for me. Every day, there were so many medications to give and we had to keep track of his every move. The severity of his medical conditions left me with a constant dread. Losing Ernie would have been devastating to me. Even thinking about it tore a little hole in my heart. Many times, I felt like he spent more time in the car, going to medical appointments, than he did running around our house.

Indeed, it was those times when he ran around, bouncing from one chair to the next, that I could relax a little and maybe even laugh at his many antics. Often, when Ernie was perched on my head, he would grab onto my hair and we would take a quick run through the house. Invariably, he would slip and end up dangling from my ear or mouth. Fortunately, he weighed less than a pound, so it wasn't uncomfortable; although, I quickly learned not to wear any jewelry since it would get torn off by a hanging monkey.

He loved to steal things from my pockets, chattering hysterically as I chased him. He would hide his treasures, repeatedly moving them if I got close to finding them. He had a true knack for taking the one item I truly needed, like my car keys. It may have been inconvenient, but I could never resist his antics. As he raced from room to room, leading me on a merry treasure hunt, I always ended up laughing.

One of his favorite games was to play "trampoline" on my bed. He could seemingly bounce up and down for hours, screaming with delight the entire time. It was a perfect bedtime activity for him, as he wore himself out and would fall asleep soon after.

And so it happened one night: After an hour of bouncing on the bed, he snuggled under my chin, clutching his favorite teddy bear, and went to sleep. Looking down at this precious little monkey, I smiled and another exhausting day drifted away. Soon after, I drifted off to sleep, too.

Suddenly, I was awake. Had it been minutes or hours that I slept? I wasn't quite sure. I looked at the clock. It was 4 a.m.; I could still get a little more sleep. I reached down to comfort Ernie and cradle him in my hand, but something was very wrong. He was ice cold and wasn't moving. I screamed, waking my husband and my daughter, who was visiting us for the weekend. Ernie was limp and lifeless. He wasn't breathing and I couldn't feel a heartbeat. I lifted his eyelids. They were fixed and dilated. I couldn't seem to move fast enough as I began CPR.

I could barely control my sobbing long enough to deliver the chest compressions and mouth-to-nose breathing. Time slowed to a crawl as 1 minute, 2 minutes, and then 5 minutes passed. I couldn't give up. My husband and daughter were watching me with tears in their eyes, not wanting me to stop, but beginning to face the horrible reality I couldn't believe.

"He's gone honey," my husband quietly said, reaching his hand out to comfort me. Then, as I was about to give up, Ernie stirred and gasped. Once. Twice. Then he stopped breathing again. Again, I began CPR. I knew I could bring him back and I wasn't giving up. Two minutes later, he started breathing again and opened his eyes, just a little—as if to say, "Have I missed something?" I looked at him, with tears of joy streaming down my face. His little face looked back at me and to my horror I saw that half of it was drooping—a sure sign of a stroke.

Hardly 15 minutes after waking up to a nightmarish scene, we were bundling Ernie into the car, having called our veterinarian and arranged to meet him at his office. There, we were able to confirm that he had suffered both a stroke and a heart attack. The emotions came rushing into me like a torrent. I had to sit down, as I was shaking and crying uncontrollably. Somehow, this fragile little monkey had beaten back tremendous odds and, through a miracle, had survived.

Yet, even with his many, many health problems, Ernie began to grow and thrive. We had devoted an entire room to him, with lots of swings, perches, teddy bears, puzzle feeders, and numerous other toys. We even built a tree house for him, right inside our house. When the weather warmed up, we constructed a huge outdoor enclosure, which he could reach through a little trap door in his room. He loved spending summer days outdoors, letting the sun warm his fur and watching the world go by. Ernie had so much space to roam and stretch. Then, every night he would come inside, wait for me to open the enclosure door, and attach himself to me for the night. It was a wonderful routine that brought me such joy.

And then Ernie started to grow up. When he was about 3 years old, he began having issues sharing his toys. I was the only person allowed to touch any of his belongings. At first it was funny and reminded me of my children when they wouldn't share. My husband, Skip, and I would make jokes about his sharing issues and we thought it was hilarious. Those funny moments were fleeting.

One night, as I lay in bed, with Ernie attached to me as he had always been, Skip rolled over to kiss me on the cheek, as he always did before we went to sleep. Suddenly, Ernie, who had been almost asleep, jumped up and began a verbal assault at Skip, chattering at the top of his little lungs. As he finished this barrage, he reached out and slapped Skip across the face. While the slap didn't hurt and we thought the incident was highly amusing, it signaled a change in our little family dynamic.

Ernie was getting bigger and more aggressive. Our sweet little infant was beginning to mature and exhibit the behaviors common to every male macaque. He was trying to exert his new-found strength and maturity to rule the troop (in this case, Skip and me).

Ernie began to make rules; rules with consequences if broken:
1. Nobody touches Polly in front of Ernie.
2. Nobody eats in front of Ernie unless they intend to share the food with Ernie.
3. Nobody eats food that Ernie doesn't like to eat.
4. Nobody touches Ernie's toys unless he first gives permission.
5. Ernie decides what objects are toys.
6. Objects of value to a human are valuable to Ernie, therefore are now Ernie's toys.

Ernie

ABOVE *On warm, sunny days, Ernie loved to be outside and feel the wind in his hair or to sit in his hammock (background).*

ABOVE *Even with all the climbing equipment and toys in his outdoor enclosure, Ernie wanted his human caretakers to come in and play with him.*

Breaking any of these rules resulted in a verbal barrage of screams and hoots, often followed by a slap across the cheek.

If I told other people of these rules, most would tell me it was ridiculous to put up with these "bad" behaviors. They'd say, "You need to discipline him and put him in his place." This well-meaning advice was so misplaced. Macaques are not dogs, who will tolerate gentle rebuke. They are not people, who can understand the explanation of why their behavior is unacceptable. Macaques are social monkeys with a social hierarchy. Both male and female macaques have to establish their place in the hierarchy. Once established, they almost always look to improve or maintain their social rank. Especially for males, that means they must challenge any other males—often quite fiercely. In the wild, this arrangement works well and promotes troop dynamics. As a pet, this behavior is almost always disastrous for the monkey (especially if the particular monkey—like Ernie—is not one to back down without a confrontation) and results in them being locked in a cage or promotes the horrible abuses I've rescued monkeys from over the years.

I knew the time had come. Ernie could no longer have the run of the house or sleep with us. We carefully designed a special monkey bed for him and attached it to the wall in his room. The bed had surrounding rails, so that if he had seizures in the middle of the night, he wouldn't fall out. His room was directly across from our bedroom, so that he was able to see us. More importantly, I was able to see him. If he had a seizure, I could quickly go into his room to hold him and make sure he didn't bite his tongue. As I held him in my arms, he would stare at me, his eyes wide with fright. Soon the seizure would be over and he would snuggle deeply into my arms, not wanting to leave their safety for hours afterwards.

Still, even with everything we provided for Ernie, he was alone. From the dozens of books and articles I had read about macaques, I knew they were social monkeys. Ernie needed a monkey companion. I was not about to get another one from the newspaper. I simply couldn't bring myself to support such a terrible pet trade, even if it was to give these monkeys a better life.

Just when I was beginning to despair about how to get a friend for Ernie, I received a phone call from a person who needed to find a home for their pet macaque. To this day, I have no idea how they found me, but I was so grateful they did.

Justin was only 6 months old, but had already been in three different homes. In each case, I was told, he had become aggressive and the owners gave him away. At the time, I knew nothing of the dynamics of pairing monkeys, but I did know that Ernie needed a companion and I felt certain that a 6-month-old macaque couldn't be aggressive. Without further ado, I arranged for Justin to be flown to Oregon, and soon I was at the airport looking into a crate at a little monkey as he "fear grimaced" back at me.

When we got Justin home, we slowly opened the crate door while speaking gently to him, aware of the reports of his aggressive nature. Sure

ABOVE *Sometimes, the best part of my day
was a conversation with Ernie.*

enough, as soon as we opened that door, Justin came flying from the back of the crate and tried to bite my hand. Without even thinking I opened my mouth slightly and scolded him a bit verbally (macaque style) as Ernie would do to me when I broke his rules. The effect was immediate, as Justin went from a little terror to a little angel, lip-smacking and cuddling against me, as if to apologize for his actions. I almost dropped him, I was laughing so hysterically.

It didn't take long for Ernie and Justin to become best friends. They played together constantly, chasing each other indoors and out, catching bugs, dipping their toes in the wading pool, and hooting at me whenever they had the chance. Everything was almost perfect, except for one very significant problem. Justin was terrified when Ernie had a seizure. The sight of Ernie quivering through a seizure would send Justin running as far away from him as possible, screaming at the top of his lungs. It became so upsetting to Justin that I had to separate them at night, when Ernie had most of his seizures.

As perfect as their friendship was, I knew I could no longer keep Justin and Ernie together. Thus began my monkey sanctuary, like my wildlife rehabilitation center—a fortunate accident that resulted in so many saved animals. Because Justin and Ernie were no longer compatible, I had to take the next steps, putting out calls to humane societies and rescue leagues to let me know of unwanted monkeys. My troop was expanding.

Why Macaques?

I'M OFTEN asked, "Why macaques?" The simple answer is, "Because the first monkey I ever brought in was a macaque." In reality, the answer is much more complex.

To begin with, macaques are fascinating monkeys. The 22 species of macaques are distributed across southern and southeastern Asia into Japan, with one species (the Barbary macaque) in North Africa and the southern tip of the Iberian Peninsula. As such, they are the second most widespread primate genus besides humans. They are almost all highly social monkeys, with some species living in troops of up to several hundred individuals. Each troop has an intricate social hierarchy. Maintaining the hierarchy is the job of the dominant male, who accomplishes this task with a look, threats, vocalization, or as a last resort, violence. Female macaques also have a hierarchy within the troop, which depends on their familial relationships. Daughters of high-ranking females tend to be high-ranking, while daughters of low-ranking females tend to be low-ranking.

The dynamics of their social hierarchy has led to macaques engaging in more overt aggression—far more so than their New World counterparts (e.g., squirrel monkeys, capuchins, and spider monkeys). As a result, few macaques are kept in zoos and very few sanctuaries will take them in.

They also share many anatomical, physiological, and even psychological traits with humans (as well as 93 percent of our genes), which has led to their extensive use for laboratory research. Many people forget that rhesus macaques gave their name to the Rh factor, in reference to a protein on the surface of red blood cells that was described with blood from rhesus monkeys. At the same time, they can harbor diseases that, on rare occasions, have been transmitted to humans through bites and scratches.

Macaques range in size from the petite 4- to 6-pound toque macaque to the 40-plus-pound Tibetan macaque. Some species have very long tails, such as the long-tailed and lion-tailed macaques, while others, such as the stump-tailed and Tibetan macaques, have virtually no tail. All macaque species are arboreal, although they can be equally at home on the ground. Unlike many other primates, macaques love to swim.

Macaques are highly intelligent omnivores. In the wild, they spend much of their time foraging for seeds, fruits, insects, and lizards, storing food in their very large cheek pouches. However, they have also adapted to living around people. Many cities in Asia have large resident macaque populations, living off scraps and handouts. This has resulted in these urban macaques being viewed as tame pets, instead of wild animals who

have adapted to city life. In some cities, the population of macaques has grown so large that they are viewed as urban pests.

In the United States, macaques are among the most commonly sold monkeys for the pet trade. Every year, thousands of baby macaques are sold to people who are ill-equipped to give them the care and life that they need. Many of these babies come from disreputable breeders who try to produce as many babies as possible, while tinkering with ill-conceived breeding schemes to create "designer" monkeys.

I knew none of this when I first purchased Ernie. I just knew that selling a baby monkey seemed like a bad idea and that I wanted to give him the best quality of life. It was only later that I discovered the vast scope of the legal and illegal pet monkey trade.

Even then I didn't start out with the idea to build a sanctuary dedicated to macaques. When I needed to find Ernie a companion, it was a fortunate coincidence that I was able to take in Justin, who was the same species. Then, in the same manner as I had created an accidental backyard wildlife rehabilitation center, two macaques became many.

By 1998, it had become apparent to me that I needed to do more. I had spent a lot of time researching other monkey sanctuaries and talking to the people who had devoted their lives to providing for the monkeys. I had found that virtually no facilities were dedicated to macaques. Between the thousands of macaques being sold as pets every year and the lack of facilities that would take in macaques, there was an enormous need. I started the Oregon Primate Rescue on my farm in Dallas, Oregon. It started very small, with the capacity for just a few monkeys; only what we were able to afford. Unlike wildlife rescues, these monkeys could never be released and would need care for the remainder of their lives (potentially over 30 years).

Word of mouth is a powerful tool and soon people all over the country were calling, hoping we would take their monkey. Initially, we focused on unwanted pet monkeys, since there were so many of them who needed help. Some came from well-intentioned people, who realized they couldn't provide the proper care for the monkey. Others came from horrific conditions, having been confiscated by local or federal authorities. If we had the room, we took the monkey, no questions asked. Later, we started receiving inquiries from research facilities that wanted to retire their monkeys. This opened a whole new segment of the monkey population, which we are only just beginning to explore.

So, when asked that question, "Why macaques?" my response is still "Because the first monkey I ever brought in was a macaque." Now, I also add, "Because there are so few options for them and I want to provide them with the best possible life for the remainder of their days."

Summer

WHEN you raise a child, you are nurturing and shaping their development. Decisions you make can last for years into the future, affecting how your child fits into human society. When monkeys are raised by their mothers, the same principles hold true. Lessons taught by their mothers help baby monkeys become a part of monkey society. Problems inevitably arise when monkey babies are raised by humans and forced to try to fit a human standard. As I had learned with Ernie, when raising a monkey it is best to continually remind oneself that you're raising a monkey, who will want to fit into a monkey society.

Like Ernie, Summer was taken from her mother when she was only a few days old. Yet, from such a similar start, their paths diverged markedly. I got Ernie as an infant and spent hours every day playing with him, provided him with an entire room to live in, and was constantly reading or talking to people about the best ways to raise a baby monkey. Summer's upbringing couldn't have been more different.

For over 17 years, Summer suffered at the hands of ignorant and misguided owners. For at least part of that time, she was kept in a tiny wire crate on top of a dresser, only coming out on rare occasions to be shown off to friends and family. The rest of her time was spent rocking neurotically under a blanket, in a scene frighteningly similar to news stories I had seen of neglected orphans in Eastern Europe. Worse, her lack of adequate room to stretch caused Summer's leg muscles to atrophy, making it almost impossible for her to fully extend her legs and move normally.

For her entire life, Summer had been forced to wear a diaper. As a baby, it was easy to change her regularly. As an adult, diaper changes became a struggle, with a predictable outcome; she was rarely changed, often going days between changes, and the diapers were always full of urine and feces. Being kept in soiled diapers much of the time caused a chronic diaper rash, permanently destroying the hair follicles under the diaper. Even worse, in an ill-conceived attempt to make diaper changing less traumatic, every one of her teeth had been pulled out.

Since she had no teeth, Summer was being fed a "special diet" by her owner, consisting of partially cooked bacon, fruit rollups, cake or doughnuts, pancakes with syrup, and baked potatoes; not a healthy diet for a monkey or a person. This was in stark contrast to what would have been her natural diet of fruit, vegetables, nuts, wild grains, and the occasional insect; or even a reasonable diet in captivity of nutritionally complete biscuits, which could have been softened with liquids. When I first met Summer, she was sitting on the floor in the owner's living room, regurgitating her food. I was told that she had done this on a daily basis for most of her life, and that "all monkeys regurgitate."

ABOVE *Summer was dressed in children's clothing on the day she was released to OPR. Her tail was kept folded into the clothing. It was hard not to cry when I saw her like this.*

The owner had contacted me and said she had recently moved to a state where it was illegal to keep monkeys as pets. She had been looking for someone to take Summer and had stumbled across our website. Further, she was now living with her son, who was apparently abusing Summer. She wanted to know if I could take her as soon as possible. We had the space, so I immediately agreed.

As I was talking with the owner, learning more about Summer's history, she climbed up on the sofa. "Get down!" her owner yelled violently. Summer quickly ran off the couch, ducking and putting her hands over her head, as if to ward off blows. She then started slapping and hitting herself in the face so intensely that she left bruises, all while grimacing in fear. Her owner paid almost no attention to this, casually mentioning that this was "normal" behavior for macaques. In her mind, Summer was playing a game. She went on to describe that the fear grimacing was actually Summer smiling as she played her "game." I sat quietly, stunned by the owner's total misinterpretation of basic macaque behavior; afraid that if I said anything she would not turn Summer over to me. Then, just when I didn't think I could be more surprised, I almost fell to the floor as she practically cried about how difficult it would be to give up the monkey she considered to be her daughter.

When the time came to give us custody of Summer, her owner put her into the travel crate, fully clothed. As we were walking out the door, she apologized for a small amount of hair on Summer's face, stating that she hadn't had a chance this week to shave her.

Shave her?

As if I wasn't shocked enough, as we were pulling out of the driveway, the owner came running after us, gesturing wildly that we had forgotten Summer's shoes (a likely reason for the blisters we had seen on her toes).

We took Summer directly to our veterinarian. It was clear that she had health issues, but we didn't yet know what they were and I wanted to make sure she did not have any obvious communicable diseases. We arrived at the veterinary clinic and sedated her so she could be examined. It was with much trepidation that we began to peel away her clothes and the very, very dirty diaper. I almost cried when I saw the razor burns on her back and tummy from being shaved. It only got worse. There were painful, seeping blisters from the constantly soiled diaper. The owner had kept a collar around her waist even though she spent most of her time in a tiny cage. The collar was too tight and the buckle had to be pried out of her skin. Her pelvic bones were protruding from malnutrition and her tail was permanently disfigured from being tucked tightly inside of a diaper. Even though she did not appear to have any communicable diseases, it was clear that Summer would have a long, arduous recovery ahead of her.

Upon getting her to OPR, our first goal was to transform her junk food diet into a healthy, macaque diet. We started with monkey biscuits softened in apple juice, which she readily ate. To make her vegetables and fruits easier to eat, we steamed them. We slowly started adding new

BELOW *Forced to wear a diaper for 17 years—often unchanged for days—Summer has permanent damage to her skin.*

proteins, including hard-boiled eggs, cooked beans with brown rice, and mealworms. Macaques love nuts, but with no teeth, Summer couldn't eat them. Instead, we gave her a little natural peanut butter. She would spend an entire afternoon dipping her fingers into a bowl of peanut butter and licking it off. Before long, her physical condition started to improve. Within a few weeks, the sheen in her fur returned and the blisters and scars slowly faded away.

Now came the much more difficult task: addressing Summer's mental health issues. Any time we tried to interact with Summer, she would start screaming and biting or hitting herself. It was heartbreaking to watch and gave me nightmares. We started by providing her with adequate space, extra enrichment, toys, stuffed animals, puzzle feeders, and access to TV. None of these seemed to help. I would sit for hours, watching her, trying to figure out how to undo the 17 years of horrendous treatment. Something had to be done to eliminate, or at least reduce, the frequency and intensity of her debilitating, self-injurious behaviors.

Typically, pairing with another monkey resolves most emotional and behavioral issues, but I couldn't safely pair her with another monkey. Any time Summer engaged in abnormal behaviors, it would cause great distress among the monkeys who could see her. Summer would start slapping her face and the other monkeys would scream and howl in distress. Further, since she had been raised in isolation for so long, she had no social skills. Just as humans must know how to interact with others— knowing when to play, when to talk, when to touch, and how to share, monkeys must be taught the same sort of social skills. Summer lacked these skills, and I was very concerned that she would be injured, or even killed, if I tried to pair her with another monkey.

Then, a breakthrough; after weeks of observation, I noticed that she spent a lot of time grooming her teddy bears and cooing at them. I thought that perhaps we could find a realistic macaque monkey doll, with lifelike skin and fur, for her to play with and groom. Did something like that even exist? I started searching the Internet and, to my surprise, found a doll that fit my idea exactly. It was supposed to be an orangutan doll, but looked much more like a monkey. My plan was to use this doll to start teaching Summer how to interact with me and with other monkeys. Due to her background, Summer did not like to be touched and was reticent to reach out to anyone. If there was any hope of addressing her emotional and behavioral issues, I had to convince her it was okay to be social.

Over many years of working with monkeys, I have learned that the initial step in developing a trusting relationship with a macaque is to spend a lot of time near them, observing yet remaining aloof. The more aloof I appear, the more they become interested in me. They are fascinated by this human who is neither afraid of them nor seems to show any signs of aggression. Yet, somehow the monkeys can sense if I am afraid or merely pretending to be indifferent, just by subtle changes in my body language. Instead of approaching me with curiosity, they will either run away in fear or become aggressive. It is almost as if we are playing some sort of dating game. I entered Summer's enclosure with the doll in my arms and sat on the floor, not making any eye contact with her. For the next 30 minutes, I sat on the floor, focused on grooming the doll, but aware that Summer was intently watching me from the highest ledge. I then quietly got up and left her enclosure.

The next morning after feeding the other macaques and doing morning chores, I returned to her enclosure, again with the doll in my arms, and repeated my doll-grooming session. This time, after about 15 minutes, Summer quietly came down from her ledge and crawled up onto my shoulder to watch me. Through my excitement, I somehow managed to remain aloof and continued to groom the doll. This was the first time that she had willingly come down to me. Even more importantly, while she was watching me, she was not trying to harm herself. I felt an exciting sense of accomplishment in a small breakthrough to gaining her trust. After a few minutes of watching me groom the doll, she started to groom herself—another first.

As I watched her from the corner of my eye, I could see that she had never had anyone show her how to properly groom herself. This was something her mother would have taught her. Summer had been taken from her mother at such an early age and grown up in such a poor environment. She knew nothing about how to take care of herself, how to interact with others, how to be a monkey. My excitement at her ability to learn was momentarily tempered by the sadness of what she had gone through to get here. That moment only increased my resolve to rehabilitate this monkey.

On the third day, Summer immediately approached me. Instead of sitting on my shoulder, she sat on my knee and we started grooming the doll together. I almost jumped up and down in excitement, but tried to remain calm and outwardly unaffected by the moment. For the next 30 minutes, she switched between grooming herself and the doll's arms, while I worked on the doll's head. I watched closely as Summer became more adept at grooming herself and the doll. After so many years of rehabilitating monkeys, I remain amazed at how quickly these intelligent beings learn new skills. As in the previous days, when she returned to her highest ledge, it was a signal that it was time for me to leave. I quietly left the enclosure, thrilled by another breakthrough.

When I returned to Summer the next day, she was waiting patiently for me. I sat down on the floor and she immediately began to groom the doll with me. As we sat there, grooming the doll and quietly talking to it, I was again surprised when Summer's tiny fingers started grooming the hair on my arm. I continued to groom the doll, not wanting to disturb this special moment. Then, after a few minutes, I cautiously reached out and started to groom Summer. I could feel her stiffen a little with uncertainty. Over the next hour, I felt her become increasingly relaxed as I groomed her. My fingers hurt and my muscles were stiff from sitting on the floor, but as long as she groomed me, I was determined to groom her. A vital part of the rehabilitation was to allow Summer to choose when to start and stop our sessions.

Our grooming sessions quickly became a favorite part of my day. Summer would watch me finish my morning chores and be waiting for me to come into her enclosure. While she still occasionally hit and bit herself, every day brought fewer of these episodes and more normal behaviors. Instead of rocking under a blanket, she was spending time exploring the outdoor enclosure and playing with the many toys in it. Instead of ignoring the other macaques, she was showing intense interest in them and beginning to communicate with them, making happy hooting and cooing noises in response to their calls. Every day, I could see a few more cracks in her shell, as she cautiously became a part of the world around her.

As her behavior improved, we still groomed the monkey doll every day, but only a few moments instead of an hour. There were so many other things that Summer wanted to do with her days. She clearly liked the doll, but given her self-destructive behaviors, I still took it with me whenever I left her enclosure. Now, after weeks of grooming sessions where Summer would groom me more than the doll, I decided it was okay to leave the doll with her. Although she had never held the doll, her actions around it were always gentle and I was certain it would help relax her when I wasn't around.

I carefully placed the doll on Summer's favorite ledge and watched as she quickly jumped up to it. Then, as I started to leave the enclosure I noticed that Summer was on the ledge, frantically screaming at the doll. This was the first time she had actually seen the entire doll, as it had always been carefully held in my arms. At first, I thought she was going to tear it apart, and then realized that she was focusing her rage on the doll's tiny cloth diaper. As she screamed, I rushed back to her. Summer was trying to rip the diaper off the doll. I pitched in to help and within seconds we had removed the diaper from the doll, throwing it violently to the floor, kicking it around the enclosure, and then stomping on it. Calm returned to the enclosure as I finished killing the diaper and threw it in the trash.

With some trepidation, once again, I placed the doll on the ledge and slowly left the enclosure, reassuring Summer the entire time. I was shaking from seeing such a violent reaction. Then, as I closed the door, I looked back and saw Summer gently cradling and cooing to the doll while touching and examining the doll's bare bottom. Her response now made sense; she was comforting and reassuring the doll. After spending those many years in a dirty diaper, Summer associated a diaper with the pain and suffering she had endured. Summer could not allow another being (even a doll) to suffer as she had.

Summer had finally exorcised her demons and could now begin a full recovery. Her resilience was amazing. Even as Summer's past slowly faded into her own deep memories, they always remained fresh for me. I could only think of the thousands of other monkeys like her, who were torn from their mothers. Who was going to save them so that they could recover as miraculously as my grooming partner, Summer?

BELOW *Summer's monkey doll helped heal her emotional wounds. After watching me groom the doll for two days, Summer overcame her suspicions and began grooming the doll with me—the first step in her emotional recovery.*

Amy

I'M NEVER quite sure how people find me. OPR keeps a fairly low profile and doesn't actively advertise to rescue monkeys. We have always taken a measured approach, only taking in monkeys when we had space and only adding enclosures when we had the funding to ensure the long-term care of the new monkeys. Yet, hardly a week goes by without someone contacting me about their monkey.

I had just finished all of the morning chores, when I received a phone call from a woman. She frantically explained to me that her pet Barbary macaque, Amy, was acting like she had broken her leg. While she didn't have a regular veterinarian, she told me that she had called several in the area and none were willing or able to treat Amy. She pleaded with me to help. Since she was local and I didn't want Amy to suffer, I offered to call my veterinarian, schedule an appointment, and call her back.

As soon as I hung up with the lady, I called my primary veterinarian. He was totally booked and couldn't see me until that evening. I was really worried, so I called my back-up veterinarian and explained the situation. He was willing to see the monkey, but only if I was there to help him handle and restrain her. With no hesitation, I agreed and was soon calling the woman back to let her know of Amy's appointment for later that afternoon. Surprisingly, rather than relief at finding a veterinarian to see Amy, she added a new twist; she started crying and said she didn't think she could get Amy into a crate by herself. Apparently, Amy was a very sweet monkey, but the owner had never put her in a crate before. Before long, I found myself agreeing to drive out to the owner's home to help her, while at the same time wondering how I always get caught up in these dramas.

The woman had provided a visually beautiful outdoor enclosure for Amy. It was spacious and about 12 feet tall. At the very top she had an igloo dog house mounted to the side, where Amy slept. There were even natural tree branches and logs leading from the ground up to the igloo. The cage appeared to be clean when I went to help her put Amy in the crate. Unlike so many other pet monkey circumstances, this one didn't look neglectful.

Within a very short time after I arrived, Amy was in the crate. All it had taken was a bit of gentle coaxing by her owner. I didn't even get into the enclosure, preferring to give advice from a distance, to avoid potentially scaring Amy (as a stranger). I couldn't get a good look at her legs, although I did notice a little limp and she was quite overweight. Without further ado, I helped the owner put Amy into her car and led her to the veterinary clinic.

Shortly after arriving at the clinic, Amy was sedated and the veterinarian was able to examine her. He first looked at her leg, remarking that it didn't look broken. Then he turned her over on her back and everyone (including the owner) gasped with horror. Her entire torso was covered with maggots, crawling in and out of the dead skin, with a large ball of maggots under each armpit! The putrid smell of gangrenous tissue filled the exam room and nearly knocked us over. I glanced at the veterinarian, who was clearly concerned and quite irate at Amy's condition. When he asked how long she had been this way and her owner responded, "It couldn't have been more than three months," he was momentarily speechless. The woman assured him that this wasn't deliberate neglect, and then told her story.

She was now working three jobs, so could only feed Amy very early in the morning or very late at night. Thus, it was always dark when she went out to the enclosure. She hadn't actually physically seen Amy in three months, relying on sounds of her chattering in the dark and eating her food as indications that she was "alive and well." She reluctantly admitted that she had not cleaned the enclosure much in the past three months and had just done a quick cleanup before I had come to her house. As she was saying this, I was contemplating how many times each day I see every one of the monkeys at OPR and how much time I spend cleaning enclosures. I was almost in shock.

It was during the long and laborious cleansing and debridement of Amy's wounds that I received a return phone call from my primary veterinarian, making sure that I had gotten Amy to another veterinarian. I thanked him for checking in and started telling him about Amy. As I described her age, species, and name, he suddenly interrupted me: "I've seen this monkey before." Time stood still as he quietly recounted our conversation from so long ago about a little monkey drinking Drano. With those few words, my initial good thoughts about Amy's life vanished. And yet, this was not my last surprise of the day.

Finally, Amy's wounds were cleaned and the dead tissue removed. She was bathed, topical medications were put on the wounds, and multiple antibiotic injections were administered. As the veterinarian was preparing the oral and topical antibiotics for her continued treatment, he carefully explained that Amy would need to be given a medicated bath daily for the next week and that she would need to be given the oral and topical antibiotics three times a day for the next three weeks.

Without blinking an eye, the woman looked at the veterinarian and told him that she was not capable of bathing Amy or giving her any medications. Then, to my shock and great surprise, she said that she was surrendering Amy to me, effective immediately.

You could have knocked me over with a feather. This topic had never once come up in our phone conversation or when I helped her get Amy into the crate. She wanted to put Amy into my car and walk away from her. I was incredulous, as my jaw hung low. She had provided Amy with a wonderful enclosure and apparently cared for her for over 20 years, yet as soon as Amy required anything more than food and water, she walked away. Then the other shoe dropped. She also had a very old capuchin monkey,

Boo Boo, who had lived with Amy for many years. She wanted me to come and get her, too; preferably that day.

How could I say "no"? How could I say "yes"?

How was it that every time I decided that we simply couldn't bring in more monkeys to OPR, another one (or two, in this case) showed up?

Fortunately, we did have one open enclosure. Typically, we would never house a macaque and a capuchin together. Macaques are so much larger and can be much more aggressive than a capuchin. However, this "odd couple" had apparently been together for years and with that kind of bond, I wouldn't dream of separating them. Thus it came to be; after a perfectly normal morning of chores, I was now driving home with two unexpected, new residents to OPR. Now I had to figure out how to explain this to my husband, Skip.

Soon, Amy and Boo Boo were safely ensconced in their new home. After living with their previous owner for many years, they would be very stressed by their new surroundings, so I went to collect some old toys and other knick-knacks from their old home. Of course, I knew that I would have to get that igloo. After all, it was Amy's favorite place to sleep.

As I clambered up a ladder to the igloo, I marveled at the two very different sides of Amy's care. Unlike so many other rescues, Amy and Boo Boo had an amazing enclosure, so clearly the owner cared for some of their needs. At the same time, when her own life became difficult, Amy and Boo Boo were relegated to the edges, only to be fed at night, never observed, and forced to live in squalid conditions. As with every monkey rescue, it can be hard not to judge people, so I try to focus instead on providing a good home for the monkeys. Most of these people aren't malicious; rather, they are horribly misinformed or in way over their heads. Their ignorance concerning monkey needs and what effort it takes to meet those needs results in the monkeys' suffering.

I got to the top of the ladder, stuck my head into the igloo to see the inside, and was almost blown to the ground by the overpowering smell of urine and feces. The floor of the igloo was caked with nearly 4 inches of waste. Not only had Amy's previous owner not seen her in three months, but she hadn't cleaned out the sleeping quarters either. It was disgusting and as I fought back my own nausea, I wondered how Amy could have slept in such horrific conditions; even as I knew that she had no other options. Either she slept on a bed of feces or outside, exposed to the elements. Neither was a good choice, but I now knew that Amy had a real attachment to the igloo. I would bring her favorite sleeping place—after a long and thorough cleansing and sanitation. At least this explained how Amy's torso got so infected.

I needed to do something else about Amy's favorite bed. She was using the igloo as a toilet. I needed to find a way to keep her bed clean. As in her previous home, we bolted the igloo high up on a wall of her enclosure. For the first few days, every morning I clambered up a ladder to clean the mess from the previous night; definitely a short-term fix. Fortunately, I soon noticed that she only relieved herself in the back of the igloo and always in the same spot. That was my "Eureka" moment. We cut a small opening in the back, small enough to prevent a head, arm, or leg from getting stuck, but large enough to allow the waste to fall to the ground. From that day forward, Amy's sleeping quarters were never soiled again, although we still had to climb up and give it a normal cleaning regularly.

Amy and Boo Boo quickly settled into their new life at OPR. They spent their days exploring their new environment and making new friends with the other residents. They particularly loved to go up and down the tunnels leading to the outdoor access. Amy was such a gentle and calm monkey, despite her great size (almost 40 pounds). Nothing intimidated her. Boo Boo, already a small monkey, looked even tinier and frailer when next to Amy. She always looked to Amy for guidance and protection and Amy was always there to help her.

Their relationship was perfectly illustrated one day when they were in the outdoor enclosure, quietly soaking up the warmth of the sun. Suddenly, the peace was shattered by Boo Boo screaming frantically, as loud as she could. Even from several buildings over, I knew something was very wrong, so I dropped everything and ran, not knowing what could possibly elicit such screams. Panting heavily, I reached the enclosure and saw Boo Boo clinging to the highest possible point, screaming at the top of her little lungs. A small snake had slithered into the enclosure. Boo Boo was terrified. Amy was her protector and had already dispatched the unwelcome trespasser by the time I had arrived.

It took another 15 minutes for Boo Boo to gain enough courage to slowly come down from her high perch. Then, in a rush of cooing and lip smacking, she ran over to Amy, who was lying on the floor, relaxing after her morning battle. For the next hour, Amy lay there as Boo Boo lovingly groomed her soft fur. With the snake encounter safely past them, they snuggled together for a well-deserved nap in the sun.

Amy was always so protective of little Boo Boo. If any other macaque made a threat face or unwanted gesture toward Boo Boo, Amy would rush over, scolding the monkey loudly and shaking the side of the enclosure. It didn't matter if the offender was in the enclosure next door or at the other end of the building. Her large size combined with the very loud outburst quickly cowed any other macaque, reducing them to submissive lip-smacking and gentle cooing. After a few minutes, everyone understood their position and life at OPR would once again be orderly.

That's how it was with Amy. One second she was so sweet. She would sit on the side of the pool in the outdoor enclosure, carefully dipping each toe into the water (unlike almost every other macaque, who would dive right in). Or, she would squeal with delight when I brought her breakfast in the morning. Then, seemingly without warning, Amy would explode, suddenly deciding that an OPR staff member, after spending months in

close contact with her, was no longer welcome in the cage. When Amy made a decision, it was best to look out.

And yet, Boo Boo was increasingly showing her age, becoming more and more infirm. Amy was so gentle with Boo Boo, bringing her bits of fruit. Every day, for months, I would come into the building, wondering if Boo Boo had made it through the night. Somehow, the ancient Boo Boo would be there, waiting for breakfast. Until one day, Amy's squeals of delight at getting breakfast didn't greet me. Without even looking, I knew that Boo Boo had died.

For weeks afterwards, Amy was a much quieter monkey. The loss of her life-long companion had clearly affected her. Instead of squeals of delight at breakfast, I was greeted with a short grunt. Instead of berating her neighbors for misbehaving, she just turned her back on them. She would lie in the sunbeams for hours, staring at something I could only guess at. Whenever I could, I would go into her enclosure and groom her. It seemed to give her some comfort and it certainly made me feel better.

Amy's loneliness and grieving affected everyone at OPR, casting a long shadow over the grounds. Just when we were beginning to consider which macaque we could pair Amy with, Rambo came into our lives. Rambo was another elderly capuchin, who had been given to us after his owner could no longer care for him. While OPR specializes in macaques, I always find it difficult to say "no" to any monkey rescue. The opportunity to improve their life is just too important to me.

Another frail, old capuchin; the symmetry of the moment was too special. Without further ado, we started the pairing process.

After years of pairing monkeys, I can comfortably say that the process is part skill, part voodoo, and a lot of patience. Some monkeys can be paired

LEFT *Even though she was very old and frail, Amy's companion, Boo Boo, loved to explore every nook and cranny in her enclosure.*

in a matter of days. Others can take weeks. Monkeys who should be compatible can fight for no apparent reason, while total "odd couples" get along famously.

Our first step was to put Amy and Rambo into adjacent enclosures. This allowed them to get to know each other, without risking any aggression. We watch interactions very closely during this period. If they ignore each other, we know they probably aren't a good match. If they spend a lot of time near each other, in their respective enclosures, we are a little more confident they can be a good pair. During the initial introductions, we spend a lot of time watching and listening. We watch how they respond to each other. We watch how they respond to strangers. We listen to how they communicate with each other. Will they use threat gestures or coo gently? As we watch and listen, we begin to learn about a pair's eventual compatibility.

From the moment Amy and Rambo laid eyes on each other, we knew it would be a good match. Rambo, who was about one-tenth Amy's size, was clearly smitten with her. Even between enclosures, he chattered and cooed at her incessantly, while trying to hand her fruit snacks. Amy gently chattered back at him, showing more energy than at any point since Boo Boo died.

After a few days, we allowed them to get close enough to touch each other, still separated by a barrier, for safety. This was standard protocol and we didn't want to rush this seemingly ideal pair. Both seemed so happy merely to be close to each other. And so, hardly a week after first introducing them, we slowly removed the barrier between their enclosures—hopeful that the pairing would work, but holding onto our nets in case it turned sour.

A couple of quick sniffs and Rambo promptly went to work, grooming Amy in all of her favorite spots.

I couldn't help smiling and shedding a few tears. Both monkeys needed a friend and now they both had one.

Even though they were best friends, they were still individuals. Rambo, due to his advanced age, spent much of his time sleeping and resting in the sunbeams. Amy, though also past her prime, still loved to sit on a ledge at the top of her enclosure; the better to survey all of the other monkeys in the building. The ledge also provided a perfect vantage point to watch movies on the flat-screen television we had installed.

Amy loved to watch TV, something Rambo never seemed interested in doing. She was very particular about what she liked to watch. Cartoons or any animated movies did not interest her (or any of the macaques). I'm not sure why they all disliked the cartoons. Perhaps it had something to do with the colors or animation. Movies with real characters, especially if they included children, could hold their attention for hours.

Of course, because the monkeys are all individuals, they each have favorite TV programs or movies. It is always easy to tell who likes a program. If they watch for a few minutes, then go off to play or go outside, I know it isn't a favorite. If they watch the entire show, chittering and hooting at the characters, I know they like it.

Over the years and hundreds of movies and TV programs, there are only two movies that have engaged every monkey, no matter how many times I showed them: *Uncle Buck,* with John Candy, and *Elf,* with Will Ferrell. They all seem to adore these two actors. It is always a great time when we pop up buckets of popcorn for all the monkeys (and ourselves) and put in one of those movies. Of course, between the impressive sounds of a roomful of monkeys crunching popcorn (they are not polite eaters) and commenting on the action in the movies, I rarely hear the dialogue. Sometimes, their responses are very intense. When John Candy was tormenting the disrespectful boy dating his niece, Amy would rush to the front of her enclosure and shake the wire violently, in apparent support of his actions. Any time the cameras zoomed in on his face, she would lip smack wildly. It was incredibly endearing and entertaining.

Sharing these happy times with the monkeys keeps me going through all of the hard work and energy it takes to run OPR and care for all of the residents.

We had watched *Uncle Buck* the night before. As always, the monkeys had been loud and raucous, Amy being the loudest of the bunch. I woke up the next morning to an alarming sound coming from the intercom system in the monkeys' building. Instead of the usual greetings, hoots, and chitterings, all I could hear were the sounds of monkeys cooing. It wasn't even a gentle cooing. It sounded so pitiful and sad. Something was very wrong.

We quickly pulled on some clothes and ran to the building. I opened the door and saw virtually every monkey hanging on the front of their enclosures. I started doing a quick visual inspection of the enclosures, and then I saw her. Amy was on her favorite sleeping ledge with Rambo right next to her. "Amy," I called. There was no response. Rambo was gently rubbing her back and occasionally pushing on her, trying to wake her, cooing pitifully the entire time. I almost fell over, crying; the entire scene was forever seared into my mind.

Amy had died in her sleep. The suddenness of her death shocked me. While she was old, she was not infirm and hadn't shown any signs of slowing down. Even the night before, she had been so full of life. Perhaps Amy had made one final decision. Through my tears, I couldn't help but smile. I would have expected nothing less.

A Day at OPR

WHENEVER I tell people that I run a monkey sanctuary, their first response usually includes "You must have so much fun playing with the monkeys." That is absolutely true. I love working with the monkeys. They enrich my life in so many ways. Not a day goes by where I don't learn something about them or myself. At the same time, there is a tremendous amount of work that has to happen every single day and it usually starts very, very early:

4:00 A.M.

I know it is 4 a.m. because just like he does every day, Cosmo, a baby long-tailed macaque who just came to OPR, is sticking his finger in my ear, while "gorilla" bouncing on my pillow. With Cosmo around, I never have to set an alarm.

4:30 A.M.

Cosmo hitches a ride on my hip as I (still in my pajamas) take him to an enclosure where he can spend the day watching the "big kid" monkeys, one of whom I hope to pair him with soon... very soon!

4:40 A.M.

As I'm about to leave his enclosure, Cosmo grabs onto my finger. He wants me to stay and play with him. I guess chores will have to wait a few minutes as I try to reassure him that I'll be back soon.

5:00 A.M.

Cosmo finally lets go of my finger. Very young monkeys, like Cosmo, usually take a few minutes before they feel comfortable enough to leave, even if it's for a day of play. Now I can start working through the ever-growing list of chores.

5:15 A.M.

I'm heading up to the "special needs" barn to let Summer fix my hair. The special needs barn is home for all of the rescues with extra medical or physical needs. Summer can't try out her hairdressing skills on the other macaques, so I have to fill in. If she doesn't groom me every day, she'll get frustrated and start biting herself, so this is an important part of my schedule. Before I get in her enclosure, I mix up her medication with some prune puree, to hide the taste. She loves licking the spoon.

5:30 A.M.

Summer's in a really good mood this morning. Instead of throwing food at the barn cat, she's cooing at her. The rest of the macaques in the barn are stretching awake and looking around with sleepy eyes, wondering when I'll fetch their breakfast. It's a rare quiet moment.

6:00 A.M.

I'd better fetch their breakfast. I go back to the house and head downstairs to make breakfast for the macaques. My volunteer, Kayla, is helping me this morning. Kayla is washing produce and I am making gluten-free blueberry pancakes for Summer, who has a gluten intolerance.

6:30 A.M.

The trays of food are prepared and placed in a large tub. Kayla carries them upstairs for me. I smell coffee brewing! I grab a cup, take a sip, and set my cup in the food tub, so I can use both hands to take the tub to the ATV we use to drive to the macaque areas.

6:45 A.M.

Oops... Kayla points out that I am still in my pajamas. Back to the house for a quick change into some clothes that I won't mind getting bleach stained. Now we can deliver breakfast.

7:00 A.M.

The macaques are greeting me with curious "hmmmmms" and friendly grunts. This is no surprise. I have their breakfast. Skip, my husband, is helping me deliver food trays and wondering, aloud, if I'll make him some breakfast.

7:30 A.M.

Everyone gets a visual inspection as we deliver breakfast. Why is Corney hanging upside down like a bat? Everyone else looks fine this morning.

8:00 A.M.

I'm doing medications this morning. Keiki is on heart medication and he gets his first. I crush his pill in a spoon and mix it with strawberry/banana yogurt. He licks it from the spoon. Next comes Jala's steroids. I have to mix hers with a half-teaspoon of my coffee, flavored with CoffeeMate fat-free vanilla creamer. Nothing else works and it has to be that specific creamer. George has an infected toe requiring antibiotics. These are new antibiotics, so I have to figure out the best way to disguise them. Keiki's yogurt isn't working for George. He won't even look at the spoon. Neither is Jala's coffee and creamer. Orange juice doesn't work. Maybe crushed bananas?

8:30 A.M.

Still trying to get George to take his flipping medicine. This is worse than when I had to give medicine to my kids. He thinks it's all a grand game. This game has gone on long enough and I'm running out of antibiotics before I even start using them, so I act like I'm leaving the barn. That seems to work. George starts screaming and waving his arms toward me. Guess who likes crushed bananas now? Of course, I have to chatter with him as he's eating the bananas.

9:00 A.M.

George is tired of hearing me talk now and I'm summarily dismissed. Next stop is the quarantine area, adjacent to the hospital. It's time to clean the enclosures for our two new research retirees, Patrice and Franco. They're just the cutest little goblins. Patrice stands up, showing me his belly; a sign he wants a treat. Franco thinks that he should get a treat, too. How can I resist?

9:30 A.M.
Time to clean Cosmo's enclosure. Fortunately, he is usually a very tidy monkey, particularly for such a mischievous baby, so it doesn't take too long. As always, he sits on my shoulder and inspects my work. The end of our routine is always the same; I kiss his sweet little toes before I leave. Skip is watching and reminds me that isn't very sanitary. I assure him that I always wash my lips before kissing any monkey toes.

10:00 A.M.
Back to the special needs barn to clean Tyler's enclosure. Tyler gets very angry and belligerent if anyone but me cleans his space. Kayla and Skip go to the in-ground pool that we have for the macaques. Yesterday, I noticed a little algae forming around the edge, so it needs to be emptied and scrubbed, a long and laborious job.

10:30 A.M.
Since Skip and Kayla are still working on the pool and I'm finished cleaning Tyler's enclosure, I start cleaning the other enclosures in the special needs barn. They'll be cleaning the pool and surrounding area for at least a couple of hours, so I'm on my own.

11:00 A.M.

I put a load of sheets, towels, and other "monkey" laundry into the washer. The washer is almost continuously running from the multiple loads of laundry the monkeys generate every day. Every time I pass the wash room, I put another load in the washer or dryer and fold the clean laundry. While I'm here, I scrub out the barn toilet, sink, and shower—not a daily chore, but I do it several times every week to keep everything as clean as possible.

11:30 A.M.

I make Kong toys for the special needs monkeys. Summer doesn't have any teeth so I smear peanut butter inside of a PVC tube that I "torched" to make it look like bamboo.

12:00 P.M.

Finally, a brief respite. Kayla, Skip, and I go back to the house, grab a sandwich, and I drink my cold coffee and creamer, at least what's left after I sacrificed it for monkey meds this morning. I run out and spend a little time playing with Cosmo and his favorite teddy bear, while Skip and Kayla are still eating.

12:30 P.M.

The three of us go to the macaque barn to clean the first five indoor enclosures. We'll do the other five after we run into town to get more produce. I gather the shop vac, ladders, scraper, pitchfork, RV brush, cleaning rags, and drying towels while Kayla puts hot water and cleaning solution into the mop bucket. While we're working on the indoor enclosures, Skip starts cleaning the attached outdoor areas. We pick up the toys, which are strewn from one end of the enclosures to the other, and put them in the bleach soak, remove the shavings from the floor, scrape and scrub the ledges, and scrub the cage furniture, walls, and chain link. Two of us can usually thoroughly clean an enclosure in about 30 minutes, if we really hustle. Today it will take a little longer. Pearly Su is showing me her toes, so I end up spending a few minutes tickling them. How can I resist those cute toes?

1:00 P.M.
Still cleaning. Eve keeps sticking her arm through the chain link and is getting really annoyed that we're not grooming her. She's slapping her wall to get my attention. I guess I'll take a 10-minute break and indulge her.

1:30 P.M.
Still cleaning. I take the wheelbarrow of dirty shavings to the field, where we compost them. On my way back, I make a quick pit stop at the special needs barn to check on Summer, who's been under the weather lately. I was going to play with her, but she's sitting in the sun with her doll, so I won't disturb her.

2:00 P.M.
Still cleaning. We just started on Jack and Pearly Su's enclosure. Pearly Su has been menstruating, so there are blood spatters everywhere. Jack loved the fresh plums we gave him yesterday, but they gave him loose

stools, adding (significantly) to the overall mess. Times like these are when I really appreciate having interns. "Kayla: It's time for your periodic evaluation. I want to make sure you've learned everything you need to know, so you'll be doing this enclosure on your own. I want to make sure you don't need any additional training."

2:30 P.M.
We just finished cleaning Jack and Pearly Su's enclosure. (Kayla didn't buy into the "periodic evaluation" comment, so we ended up doing it together.) Now it's time to run into town to get a load of produce. Before we leave, I put on a movie for the monkeys to watch. It's one of their favorites: a series of videos I recorded of them running around their enclosures and having a grand time. Whenever I show this movie, they are enthralled and can hardly tear themselves away from watching the screen.

3:00 P.M.
We're in the produce aisle of WinCo with two shopping carts. I've gotten really good at this and can fill both carts to the top in less than 30 minutes. We get stopped often and asked by other shoppers if we have a big family and how can we eat 20 bunches of bananas? Since I'm always rushing, I usually agree that we do have a large family that loves bananas. I just "forget" to tell them my family is a troop of macaques.

4:00 P.M.
We're back at OPR, unloading the food and putting everything away. It always looks so colorful in the monkeys' kitchen on produce-shopping day.

4:30 P.M.

Now that all of the produce is put away, it's time to start making everyone their dinner. I'm boiling rice noodles for Summer to go with her steamed vegetables and fresh fruit. Everyone gets a tamarind tonight—one of their favorite treats. They love to peel away the outer pod and chew the fruit right off the seeds.

4:45 P.M.

I feed Cosmo and the boys in the quarantine area. I love the low, grumbling, happy noises they always make at feeding time. The crunching sound of monkeys chewing carrots is music to my ears.

5:00 P.M.

Summer gets her meds and then we feed everyone in the special needs barn. I also feed the barn cats. Summer and Tyler throw fits, screaming and banging around their enclosures, when I reach down to pet our orange cat, Spaz. It's been a long day and I forgot that both of these monkeys get very upset if I touch anyone else (animal or person) in front of them.

5:30 P.M.

Feeding time in the main barn. Even though I'm hungry and tired, this usually takes only 30 minutes. Everyone comes to greet me except Jack. This is highly unusual, as he and Pearly Su are always together, so I go into their enclosure to check on him. Oh no!!! Jack has his finger stuck in the water dispenser. He's very upset about this, so we will have to be careful to avoid getting bitten or scratched or damaging his finger.

5:45 P.M.

Still trying to free Jack's finger. This is the first time any monkey at OPR has ever gotten their finger stuck in a water dispenser and it's distressing to all of us. Just when I thought that I had accounted for all of the ways the monkeys could get themselves in a jam, Jack finds a new way. Each macaque is a unique individual and has unique talents, interests, and abilities to get into trouble.

5:50 P.M.

It's apparent that we won't be able to easily extricate Jack's finger from the water dispenser without injuring it. So, we lock Pearly Su outside and sedate Jack.

6:00 P.M.

Now that Jack is sedated, we're able to lubricate his finger with a little Vaseline and remove it from the water dispenser. Thankfully, it appears intact and uninjured. Hopefully, he won't try to stick his finger in there again. Of course, we'll have to figure out how to modify the dispenser to prevent any recurrence.

6:05 P.M.

We put Jack into a pet carrier on the floor of the enclosure. Typically, the effects of the light sedative wear off in about 30 minutes. Until then, Jack will be a bit wobbly and we don't want him to climb to a ledge and risk falling. I'll hang out in the barn and wait for him to wake up.

6:10 P.M.

Lunch was a long time ago and I'm starving. As I wait for Jack to wake up, I eat some peanuts from the monkeys' trail mix and spend a little time playing with Holly.

6:15 P.M.

I check on Jack, who should be starting to wake up. I get a big surprise, when I see Jack staring at me from the front of the enclosure. Apparently, he had decided it was time to leave the pet crate and let himself out. The pet carrier is in pieces. As I look around and gather up the parts, I can see that he must have pushed the top of the carrier apart just enough to allow him to lever the door open. From there, he quickly dismantled the entire carrier. I'm trying to put it back together and realize that I'm missing a spring. Jack is still a bit groggy, but is watching me closely. I look at him and realize that he's holding the spring in his hand. He's not in a giving mood, so I have to trade him a slice of fresh pineapple for the spring.

6:30 P.M.

Now that Jack has woken up a bit more, I open the guillotine door and let Pearly Su back inside. She immediately runs over to Jack, hugging and grooming him as they chitter back and forth.

7:00 P.M.

I turn on some Indian flute music at a low volume and say "good night" to my friends. The music will automatically turn off in about an hour.

7:15 P.M.

A quick stop in the quarantine area, on my way back to the house. I feed Cosmo and the boys a light after-dinner meal. Cosmo made a royal mess today. There are bits of his (formerly favorite) teddy bear strewn across the enclosure. Fortunately, that can wait for the morning.

7:30 P.M.

We were going to go out for dinner, but it's been an exhausting day and no one has any energy. I guess it's soup and sandwiches.

8:00 P.M.

I fetch Cosmo from his playground enclosure and let him swim in the bathtub.

8:30 P.M.

The royal mess followed Cosmo into the bathtub. I'm now drying the walls, ceiling, and floor of the bathroom. Somehow the bar of soap ended up in the toilet. Cosmo is so proud of himself, but he's played hard all day and is really sleepy. After I finish drying the bathroom, I towel him off. He promptly latches onto my neck and snuggles in.

9:00 P.M.

It's bedtime. Cosmo is attached to my neck and mumbling in his sleep. As my head hits the pillow, I can't stop thinking about Jack getting his finger stuck. He has access to nearly 20 water dispensers. Am I going to have to modify every one of them? What about Summer? Will she be feeling better tomorrow? Will Cosmo let me sleep a little longer... maybe until 4:30?

George

NOT ALL baby monkeys come from breeders and dealers. There is a
far darker and more appalling black market trade in wild-caught baby
monkeys. When monkeys are taken from their natural homes, they almost
always leave behind dead families. The trauma experienced by these
infants is terrible. The conditions they must endure as they are smuggled
into the country are unimaginable. The infants are often permanently
scarred, physically and emotionally. I often wonder if the people who buy
these tormented infants know, or care, how the monkey got to them.
Would they be so quick to buy a wild animal? Would they condemn them
to a solitary life in a cage? George's rescue was a miracle, but was only the
first step in a very long recovery.

It was a chilly December day in 2007. Like most winter days in the
Northwest, a blanket of gray mist covered the landscape. Sometimes, the
mist is comforting, keeping everything tucked in and safe. Other times, it
can be dark and ominous. As I looked at the mist from my kitchen window,
the telephone rang. A federal wildlife agent, who knew of my sanctuary,
wanted to know if I had room to shelter an infant rhesus macaque who

had been smuggled into the United States from Thailand. As the mist became increasingly dark, I listened to George's heartbreaking saga.

George was abducted from his birthplace in northern Thailand when he was only a few days old. We have no idea what happened to his mother or the rest of his troop. As macaques are fiercely protective of troop members, it is likely that the entire troop was killed so a single baby could be taken. I could scarcely imagine the terrifying scene that George must have witnessed. This is the sad fact of the illegal wildlife trade.

Having just wrenched him from his family, the traffickers had to figure out how to smuggle him from Thailand and into the United States. Over the next few weeks, they repeatedly drugged him to determine the right combination of anesthetics necessary to keep him asleep for the grueling 20 hours it would take to get from Thailand to Los Angeles and then through customs. Finally, on the day he was to be brought out of Thailand, the drugs were administered and George was stuffed in a pouch strapped to the waist of one of the smugglers—who wore a maternity blouse to complete the disguise. The masquerade worked exactly as planned, as the traffickers passed through Thailand customs and boarded the plane undetected.

Somewhere over the Pacific Ocean, George began to stir. Rather than risk having him wake up, the smuggler ran to the bathroom and injected George with another dose of drugs. It is amazing that he didn't die from a drug overdose, or suffocation, or starvation. I can only wonder how the other passengers would have responded, had they known the terrible events that were occurring right next to them.

After the airplane landed in Los Angeles, there was another missed opportunity to find George, as the smuggler's luggage was randomly selected for a more thorough search. Unfortunately, the customs officials never suspected that there was a baby monkey hidden under the maternity blouse, and soon George was in a rented car for the 1,200 mile trip to the smugglers' home in Spokane, Washington. Again, I marvel that he survived this third, grueling trek.

And then, just when they believed they were safe, the traffickers made the error that led to George's rescue and their incarceration; they took this helpless baby monkey out in public and bragged about how they "acquired" him. Thankfully, a concerned store clerk who heard their story followed them into the parking lot, wrote down their license plate number, and called the police. Soon, these callous smugglers were arrested and George was rescued by federal Fish and Wildlife personnel.[1] Thus began another journey for George, eventually leading him to me.

With George's rescue, the federal agents faced a serious problem. What were they going to do with a helpless, 4-week-old monkey who was smuggled into the country? George's first stop, after being rescued, was to

---

[1] *The smugglers were tried and convicted of smuggling and conspiracy.* www.fws.gov/news/ShowNews.cfm?ID=EF38B798-FE9C-34D7-FA635FBC6DE29EE1

be put in strict quarantine at a nearby research facility, as he could have been carrying a number of communicable diseases (both for monkeys and humans). For the next three months, he was isolated in this facility, so he could be observed and undergo the extensive testing necessary to ensure he was in good health and free of risk to others.

Thus was the cruel twist of fate for George. He had been rescued from the clutches of the smugglers, but was not yet free. At his age, George should have still been cuddling in his mother's arms, suckling, and being groomed. He would have been just old enough to start playing with the other babies in his troop, running through the forest, and learning social skills. Instead, he was alone in a sterile cage, in a barren environment, with no other monkeys to comfort him when he was lonely or scared.

I cannot remember how long I was on the telephone with the agent as he told me George's sad story. Finally, he hesitantly asked if I would be willing to take George once he was released from quarantine, while the case against the smugglers worked its way through the legal system. He told me the case could take several years and that because George had been illegally brought into the country, the court might eventually order that he be euthanized. As I struggled with this harsh reality, I was told that I needed to keep everything about the monkey strictly confidential. Absolutely no details about him could be made public or the case against the smugglers could be lost and they would receive no punishment for their horrible crime.

It was a terrible choice, but I knew that I had to take him in. I would bottle feed, nurture, and love him. I would develop a deep connection with this little monkey. I would try to help him recover from the trauma of his first months of life. Then, I would hope that the court would show mercy for this tiny victim of senseless cruelty—and mercy to me, because I knew that if the court ordered his death (which, I was told, was about a 90 percent likelihood), I would be completely crushed. How could I possibly put myself in such a painful position? I hadn't even seen a photograph of George, yet my heart was already aching as I thought about what might possibly happen in the future.

"George?"

That wasn't even his "real" name. Up until that very moment he was actually called "Apoo." Yes, I had already changed his identity in my mind! His name would be "George" as in "George of the Jungle." With that decision, final arrangements were made for him to be transferred to the OPR facility immediately following a three-month quarantine.

Then, on an early spring day as the sun peeked through the mist, George arrived. There he was, so tiny and curious as he peered out of his transport crate. Knowing that he had not been handled for the duration of his quarantine, I put on a pair of leather gloves as protection in case he tried to bite me, out of fear.

My precautions were totally unnecessary. The moment I opened the crate, a tiny bundle of joy peered out at me. We looked at each other for only a moment and then George jumped into my arms and clutched at my blouse, cooing and snuggling in for a nap. Somehow, in that single instant when we first looked at each other, George knew he could trust me and accepted me into his life. Even after 10 years of rescuing monkeys, nothing could compare to the emotion of that single moment. I cried. Okay, I thought, this was going to be really hard—knowing that my heart had already melted into a heap over this innocent baby and he had been here less than 10 minutes.

As I took off the gloves (never to use them again with this monkey) the sun was bright and I began the incredible task of raising George. My mind raced as I began planning his life at the sanctuary. He would have a heated, in-ground pool to splash in, trees to climb on, a place to chase butterflies and frogs. He would be introduced to other macaques and build the friendships he had been denied by the heartless smugglers. I was breathless with anticipation, even as I tried to forget about what might happen in court.

In the beginning, we had to keep George away from the other macaques. Even though he had been through one quarantine period, we decided that given the circumstances of his life to this point, it would be best to isolate him for another three months. While not ideal, we would use this period to begin nurturing him and developing his social skills.

In addition to the social skills that George needed to be taught, he also had a very significant physical malady. Due to the drugs used when he was abducted, George's tiny hands shook almost uncontrollably, making it very difficult for him to grasp onto anything or even feed himself. Although we hoped that he would outgrow the trembling, we knew that while he was so helpless George would have to live with us. We would be his surrogate parents.

Since George couldn't yet be introduced to other monkeys, my husband and I took on the task of teaching him everything we could about being a monkey. During the days, we played outdoors with George, teaching him to swim and climb trees. We would laugh hysterically as he chased bugs in the grass. We allowed him to see other monkeys, although only at a distance. We still had to keep him quarantined from the others.

Nighttime was the most difficult for George and us. Baby monkeys have a need to maintain physical contact with other warm-bodied members of their family. George had been ripped from his family in Thailand and had been deprived of this contact for the past four months. Initially, George stayed in a little bed in our house, with a little doll to hug. After several sleepless nights, listening to George cry in his sleep, as he reached for a mother who was not there, we brought George in to sleep in our bed with us—not ideal for him or us, but we didn't have many options.

The change in George was almost immediate. He slept on a pillow between Skip and me. Instead of being kept awake by night terrors, we were treated to the gentle sound of George cooing as he dreamed. Some nights, he would wake up and quietly go over to me or Skip and with his

tiny fingers gently pry our eyelids apart, as if to make sure we were still there. One night, as I lay in bed, I rolled over to see Skip smiling in his sleep. As I glanced up at George, I almost burst out laughing; he had the same smile as he slept. Not wanting to let this magical moment pass unrecorded, I carefully reached for my camera. As the flash went off, both stirred and then quickly went back to their pleasant dream. As I nodded off to sleep, my heart was filled with happiness that George had been rescued and was becoming an integral part of our life.

Spring turned into summer and then to fall. Soon we were being greeted by our seasonal blanket of mist. George was growing rapidly and becoming a self-assured juvenile. As I looked out over the sanctuary and into the fields beyond, the mist was a cozy blanket over the landscape. I was interrupted by the sound of the telephone.

It had been months, but I instantly recognized the voice of the federal agent who had brought George into our lives. As soon as I heard his voice I experienced an intense panic like I have never experienced before. This was the call we had been dreading for such a long time. Were they going to take George? So many terrible thoughts raced through my mind and

LEFT TOP *George and Skip relax and take in a movie at the end of a long day.*

LEFT BOTTOM *Fast asleep, apparently sharing the same happy dream.*

then I heard the words I had not dared to expect: "He's yours." The agent (who I am sure had a hand in the court's decision) was almost as elated as me as he told me that we had been ordered full, unconditional, and permanent custody of George!

I don't think I have ever been so relieved about anything in my life. It felt like a huge black cloud had been lifted from over my head and I could feel the sun shining even on this misty day! It was truly a day to celebrate. George's life had been touched by the best and worst of humanity. Now I was free to introduce him to the other macaques. My emotions swirled as I planned out his future at the sanctuary. George had suffered tremendously at the hands of the wildlife traffickers. He could never be returned to the wild where he was born. Instead, we would provide him with the best possible opportunities to live a normal life among his captive peers.

ABOVE *Nothing beats a dip in the pool on
a warm summer day.*

Annie

IT IS A common theme with baby monkeys. People buy them because they are incredibly cute and entertaining. They rarely consider the fact that an infant monkey requires nearly constant care and nurturing. In their natural habitat, baby macaques will stay with their mothers for at least a year, learning social skills, practicing climbing, playing with other babies, and being carried around when they're tired. Too often, as I had found with Ernie and Summer, the scenario is very different for infant monkeys in the pet trade. They are taken from their mothers very soon after birth; certainly a terrifying experience for both mother and baby. They are often deprived of the care and nurturing they need. In the end, these babies rarely develop into normal monkeys.

From the moment she was born, Annie had a difficult life. Much like Ernie, Annie was the outcome of a senseless inbreeding scheme to create "pocket Javas"—miniature long-tailed macaques. Annie's mother died while giving birth to her. Not expecting to have to care for an infant monkey, her owner contacted me and asked if I would be able to take her.

I immediately answered, "Of course."

To which she responded, "... when I find the time to ship her."

For the next four months, I waited every day for the phone call telling me that Annie was on her way. As I waited, Annie lived a very lonely life, devoid of the attention, nurturing, and nutrition that every infant monkey needs. Rather than being comforted as she ate, Annie had to figure out how to drink from a bottle propped on the top of her cage. Even though her cage was large enough and she had clean blankets and toys, she was deprived of any nurturing, causing her to develop horrifying behaviors, including self-mutilating her fingers or toes and tearing apart her stuffed dolls. We were told that she vomited constantly.

Finally, after four months of waiting, Annie came to OPR. She was pitifully tiny and underdeveloped. I immediately lavished her with the constant attention that she so desperately needed. I held her during every feeding, and then burped her with gentle rubbing and pats on her back. I allowed her to fall asleep in my arms after her feedings and then, not wanting to disturb the peaceful way she was sleeping, I made a small baby sling out of soft fleece material so she could snuggle down with her teddy bear. She was so comfortable and secure there. Then, as her natural mother would have done, I kept her with me throughout the day. Occasionally she would come out of her pouch and crawl all over my body, like I was a moving tree. I was the mother she never had; feeding, nurturing, and protecting her as she grew.

As she grew older, she started creating games. She would hide in her pouch, then leap straight up to my face and try to grab onto my chin; all while making the happiest chattering noises. Playing along with her game, I would act surprised and gently wrestle with her, just as another young macaque would. Her squeals of joy always brought me to tears of laughter.

Another favorite was playing catch. I would sit on one end of the sofa, while Annie was on the other. I would gently toss a ball to her. Annie would grab it out of the air and sling it back at me as hard as she could throw. This would go on until I missed the catch, instead receiving a ball to the head. Her chittering and bouncing up and down convinced me that she thought this was hilarious.

We also created a variation of catch, with cranberries. Annie would sit on the edge of the couch and implore me to toss her a cranberry. She was amazingly agile. No matter where I lobbed the cranberry, she would catch it, quickly storing it in her cheek pouch and waiting for another. If I didn't throw one soon enough, she would slap the floor, demanding my attention. Then, one day after playing "cranberry catch," I turned my attention away from Annie for a minute. Suddenly, a cranberry hit me in the head and Annie was chittering with delight. Chuckling, I tossed it back to her, only to have it thrown back to me. Back and forth we went. From then on, "cranberry catch" was our favorite interactive game and I was forever chasing cranberries on the living room floor.

After she was weaned, I repeatedly tried to pair her with other juvenile monkeys. Annie needed social skills that could only be learned by being with other monkeys. Unfortunately, every time I tried, Annie would fly into a rage, both at me and the other monkey. It rapidly became clear to me that pairing her with another monkey was not going to be possible and would only result in serious injury to her or the other monkey.

Yet, Annie needed a companion. In the past, we had used other animal species as companions when we didn't have a compatible match for a monkey. Our first attempt was with Spaz, a beautiful little kitten we had recently adopted. The initial pairing went well. Annie was completely fascinated. She approached and then backed away from Spaz from every possible angle, trying to figure out this little ball of fluff. Finally gathering up all of her courage, she cautiously stuck out her hand, only to have it playfully batted away.

Such a blatant attack! Annie couldn't get far enough away, climbing as high as she could, all while screaming her disdain at her prospective playmate. Since we were closely monitoring the attempted pairing, we were already in the enclosure as Annie started screaming, and quickly scooped up the kitten and got her away from Annie, who calmed down as soon as her "attacker" had left the enclosure. Like most cats, Spaz was completely unimpressed with the experience. Her days would now be devoted to chasing leaves and lying in the sun. After the failed kitten trial, we tried a puppy. The puppy was from a close neighbor, who agreed to take him back if he wasn't a good match. He wasn't. Then, we tried two different rabbits. They were quickly dismissed and spent the rest of their lives being loved and cared for by my granddaughter. In each case, Annie was initially curious, but ultimately was too aggressive or uninterested.

Thus it came to be that Annie was my constant companion—my sidekick and my buddy. Much of her time was spent on my shoulder, tagging along during chores. When I poured my first cup of coffee in the morning, Annie was always there watching. As I added the creamer, her little face would sidle up against the carton, watching closely as the creamer cascaded into the coffee. Every time I stopped pouring, she would reach out a little finger to the carton and tip just a little more in. She then watched me intently, waiting for me to take that first sip. As soon as I did, she would reach up and try to pry my lips apart, as if to see where the coffee had gone.

It wasn't just the coffee. If I took a bite of food, she wanted to watch me chew. Chewing with my mouth closed was never an option with Annie around. As I chewed my food, she would imitate me, chewing on real or imaginary pieces of food. There were many days when the routine nearly brought me to hysterics. It was everything I could do not to burst out laughing, knowing that it would frighten her.

Annie's fascination with my mouth didn't end after I left the table. Every morning, she sat on the bathroom counter and watched me brush my teeth. As with the coffee and food, her face would be mere inches from mine, intently scrutinizing my motions. When I opened my mouth to brush, she opened hers, imitating my facial expressions and studying each movement. With a little practice, she learned how to turn the knob. After a few months, Annie knew exactly when I was about to turn on the water, and would beat me to it. Soon, it became her job—not mine. If I even motioned to turn the knob, I would be "scolded" verbally or have my hand slapped. With monkeys, when they take ownership of an object or a task, it belongs to them. (If they want it, it belongs to them. If you want it, it belongs to them. If you did it and they wanted to do it, you'll get "corrected.") And so, for the rest of our life together, Annie was my devoted "knob turner."

Annie observed my every move in the bathroom mirror. She watched me wash my face, floss my teeth, apply my makeup, and fix my hair. Whenever I stood with my face close to the mirror, while applying mascara, or checking my teeth, she would put her face very close to the mirror and continue watching both me and my mirror image. She would help me examine my teeth and then examine her own, using the mirror as a tool to help her find any defects or particles of food. She would use the index finger only on each hand, one pulling her top lip up and the other pulling the bottom lip down so she could get a really close and thorough view of her mouth and teeth. If she saw particles of food between or on her teeth, she would use the mirror to guide her tiny finger to remove the particles.

What Annie lacked in size she more than made up with in her ability to apply what she had observed. One day as I sat watching TV, with Annie

ABOVE *Annie delighted in copying my actions at the bathroom sink. If I looked at my teeth, she looked at hers. If I moved close to the mirror, so did she.*

perched on the sofa behind my head, she nonchalantly reached down and pulled out several strands of my hair. I pivoted around. Why had she pulled my hair?

I soon had my answer. After months of watching me floss my teeth, Annie was now using strands of my hair to floss hers. She ran them back and forth between her teeth and then examined the hair for food particles. From then on, Annie had meticulously clean teeth and my head had thin patches of hair. Occasionally, she would use a thread of bark from the branches in her enclosure (and even regular dental floss), but my hair was definitely her preferred floss.

When Annie was about 3 years old, a neighborhood boy brought me a guinea pig he could no longer keep. He knew we operated a primate sanctuary and thought we might accept his sweet guinea pig. I agreed to take the guinea pig, planning to let my granddaughter keep him. Then the most surprising thing happened. Annie showed intense curiosity toward him. It had been years since we had last tried to give Annie a companion, but her interest was so great that we decided to arrange an introduction.

It was an immediate success. Annie was incredibly gentle with the little guinea pig; we promptly named him (with no great imagination) "Annie's Piggy." Over the next few weeks, the pair were inseparable. Piggy would wander the floor of Annie's enclosure, looking for bits of fruits and vegetables. Annie would watch him eat with the same rapt attention that she had reserved for my teeth brushing, putting her head on the ground as

Piggy chewed. After eating, Piggy would sit still as Annie groomed his coarse fur, whistling and chirping with delight. It was such an endearing scene.

Then, just as I allowed myself to think about how much Annie's social skills had improved, crisis struck. I woke up one morning, went in to feed Annie, and found myself staring at her jaw. There was an enormous swelling along the jaw line, which was starting to wrap around her neck. I hurriedly called our veterinarian and within minutes we were on our way. From the time I first saw Annie until we arrived at the veterinarian's office couldn't have been more than 30 minutes, but the swelling was growing rapidly and Annie was having difficulty breathing.

We rushed in; Annie was gasping for breath as she gave me a pleading look to help her. I was in tears as the veterinarian and technicians hurriedly gave her oxygen. Slowly, her gasping stopped. We all breathed a sigh of relief as Annie received antihistamines and steroids to treat a near-fatal allergic reaction. Her life had been saved and I began crying, again.

Within a very short time, Annie had fully recovered and the swelling around her neck disappeared. What had caused such a severe reaction? We were instructed to remove everything new to her diet and environment. As we wracked our brains, trying to figure out what was new, we kept coming back to the one new addition, her beloved Piggy. Could Piggy have been the cause of this near-death experience? We had successfully introduced so many monkeys to rabbits and never experienced any kind of allergic reaction. With a heavy heart, we removed

Piggy from the enclosure and carefully washed down the floor to remove any trace of guinea pig.

Over the next week, we watched Annie closely, hoping that something else had caused her allergic reaction, but she was perfectly normal. Then, in a carefully monitored meeting, with the veterinarian present, we allowed Annie and Piggy back together. Annie was joyful as she playfully groomed Piggy and then watched him chew a carrot. We couldn't help smiling. But our smiles quickly faded as Annie began to get red splotches on her hands and face. We hurriedly gave Annie antihistamines and removed Piggy. As I watched her splotches fade, my hopes of finding an animal friend for Annie also faded. I realized that Annie's companionship was going to have to come from me. Piggy soon ended up living at my granddaughter's house, forever chasing the rabbits around their pen.

Allergies as severe as Annie's are very rare in monkeys. In Annie's case, she had the misfortune to have been highly inbred and denied the benefit of her mother's milk. Annie suffered because of someone's greed and ignorance. None of this was necessary. Annie suffered lifelong physical problems from the effects of inbreeding. She suffered lifelong emotional problems because her original owner wouldn't properly care for her. Thankfully, we were now able to provide her the care and attention that she deserved, although we would be constantly on guard for the next allergic reaction.

Ivan

WHEN I started OPR, my goal was to provide a lifelong sanctuary to the many, many pet monkeys who had suffered at the hands of misguided or heartless owners. There was such a great need to care for these monkeys, I was never able to focus on the fact that there was another segment of the captive monkey population that also needed a sanctuary: monkeys who had been in research studies.

My initial conversation with someone from the research facility where Ivan was kept was typical of those inquiries. One day, the telephone rang and a tentative voice asked, "Would you be interested in taking in an older, male rhesus macaque, who was no longer being used for research?"

My first response was an emphatic "YES!" Then came the questions.

"What research had he been used for?"

"We can't tell you."

"Is he healthy?"

"He is in good health for a 20-year-old rhesus macaque."

"Can I say where he came from?"

"We'd prefer you didn't."

"What can I say about him?"

"Well... this is new to us, too. We've never retired a monkey before, but Ivan is a real character and we don't want to euthanize him just because he is no longer needed. It would probably be best to say that he has been retired from research and leave it at that."

"That sounds reasonable to me. My goal is to provide him with a good quality of life for the rest of his days."

Thus began an excruciatingly long, paperwork-filled process to bring Ivan to OPR. First, the attending veterinarian from the research facility visited OPR to carefully inspect our facilities and be assured that Ivan would have appropriate accommodations. Then came the nondisclosure agreements, contractual agreements, health certificates, and seemingly never-ending procession of paperwork. Ivan also needed physical examinations and multiple laboratory tests to confirm his health status and assure us that he was not a risk to the other macaques at OPR.

Finally, the day came. I had spent most of the morning on the phone with the technician who had cared for Ivan for over 17 years. I could hear his anticipation, concern, and yet great hope that Ivan, a veteran of research and a favorite in the lab where he had lived his entire life, would thrive in his newfound freedom. We were both excited for his retirement, yet concerned at the same time. Ivan didn't even know the outdoors existed, what the sky looked like, what a bird looked like, or what the warm sunshine and summer breezes might feel like caressing the fur on his back.

As I drove to the airport to pick up Ivan, I was absolutely thrilled and hopeful that he would enjoy his new life and thrive in the environment we had prepared for him at the sanctuary. The indoor enclosures alone at OPR were enormously larger than the cages where Ivan had lived his entire life. In research, monkeys of Ivan's size are kept in cages with at least 6 square feet of floor space and enough cage height to allow them to stand up to full height (about 2½ to 3 feet), although many facilities will use somewhat larger cages. At OPR, the standard indoor enclosure has 144 square feet of floor space and is 8 feet high. Adding to that, the outdoor enclosures, which are accessed by tunnels from the indoor enclosures, are about the same size as the indoor ones. Then, just in case the monkeys get bored, we have two play areas that measure 900 and 2,400 square feet, respectively. All of the enclosures and play areas are filled with myriad ledges, perches, swings, and climbing branches. Instead of an unchanging view from his laboratory cage, Ivan's new enclosure featured opportunities to see a revolving landscape of horses in the meadows, deer and rabbits grazing on the edge of the forest, birds flying overhead, clouds, sun, and rain.

Could he adjust to such a drastic change after 20 years? I had no idea what to expect.

Transportation of animals is always stressful for them, and it must have been even more so for Ivan, after a lifetime in the laboratory. I could only imagine the emotions he felt while being put into a crate and loaded into the cargo hold of an airplane, separated from everything and everyone he knew. Working with macaques for so many years, I knew that the relationships they developed with one another were no different than the friendships humans develop with other humans. Yet, through my concern over the stress of his travels and the loss of his friends, I knew that Ivan would get to see and experience things that would have been beyond his wildest dreams up to that point.

My goal was to get Ivan off the plane, into our transport van, and to the sanctuary in the shortest time possible, so we arrived long before the plane was scheduled to arrive. We parked in an area where we could see aircraft as they approached and landed, as I wanted to document Ivan's arrival with a camera and a video recorder. When I saw Ivan's plane approaching, I quickly ran over to and climbed onto the chain-link fence surrounding the airport. I had a perfect view of the runway and began shooting video.

In the excitement of the moment, I completely missed the large sign stating, "Restricted Area." I was so focused on making sure the plane never left my viewfinder, I did not even hear the vehicle come screaming up to me. Suddenly, my focus was interrupted as I heard a loud, booming voice shout, "Step away from the fence! Step away from the fence!" I turned to see two security officers running toward me. I climbed down and quickly produced identification, while trying to explain what I was doing. The airport security did not share my excitement. I had to move that instant or I would be arrested. At least they didn't confiscate my camera, so I was able to obtain great video footage of Ivan's arrival.

When Ivan's crate entered the pickup area, I could see his little eyes peering through the wire mesh, intently observing all the commotion around him. I was overcome with relief. He was on the ground and he was safe! We soon had his crate loaded and secured in our van. I sat with him on the trip back to OPR, talking softly and feeding him a few peanuts. Even though he took the peanuts, it was very clear that he was nervous. He never made a sound the entire trip as his gaze shifted from one spot to the next at the slightest bump, or noise, or light.

In the research setting, people who work with monkeys are often completely covered with protective garments. This experience showed, as Ivan was careful not to touch my fingers and did not allow mine to touch his. Yet, his willingness to carefully take the peanuts from me demonstrated that he had been treated kindly in his past and he was going to adapt very well to sanctuary life.

Then we arrived. Everyone was so excited to see Ivan. As we opened the door of his crate, we were really opening the door to Ivan's new life. At first, Ivan just stared, overwhelmed by the new sights and sounds. Wonderful ledges, swings, and perches donned his enclosure, yet he remained on the floor. I wondered at the time if he had a neurological or other problem affecting his mobility. Only much later did it dawn on me that Ivan had never had the opportunity to climb up so high—he didn't realize he could!

As I watched him tentatively make his way up each level of his cage, I was like a proud parent watching a baby's first steps. I found myself quietly cheering him on with each step up.

Fortunately, we had designed Ivan's enclosure so he could gradually become exposed to the many new experiences. I could see his anxiety with the large windows to the outside world, so we quickly covered them with sheets. He calmed down almost immediately. It was weeks before we could begin gradually lowering the sheets on his windows, so that he could view horses and deer romping through the pastures, and rabbits passing by. Each new sight brought on fresh anxiety; even the beam of sunlight startled him. Little by little we were able to expose him to the new and exciting views from his window.

Several months had passed since Ivan's arrival. Every day, his unease at the sights and sounds around him decreased, yet he continued to be very apprehensive around people. He would wait until after I left before taking the treats I had put on a ledge of his enclosure. Still, I could tell that he was becoming more familiar with me. Instead of hiding at the top of his enclosure, he began to come down lower. He stayed out of reach, but I knew that he was no longer fearful.

Then, one day, to my complete surprise, as I was putting the treats on the ledge, he reached out and gently held my hand. As the hair on the back of my neck stood on end, I remained still and began quietly talking with him. He then began grooming my hand and lip smacking, his friendly invitation to groom. My heart just melted! Finally, Ivan was coming out of his shell.

Our friendship continued to grow over the next few weeks and I began to understand Ivan's moods, behaviors, likes, and dislikes. Even though I spent many hours interacting with Ivan through the safety of the enclosure bars that separated us, it wasn't enough. Macaques are social monkeys and Ivan needed someone to share his life with; a warm body to snuggle with at night and keep him company throughout the day, but we didn't have another macaque compatible with him. In the past, we have paired rabbits with macaques, as temporary companions. With no other options, I decided to try this pairing technique with Ivan until we could pair him with another macaque. This brought "Maynard Rabbito" into our sanctuary. I felt certain that Maynard–a large rabbit with a very relaxed disposition–would be a perfect match for Ivan.

Our introduction process with rabbit/monkey pairings is rather lengthy, as safety of the rabbits is equally important as the safety of our monkeys. At first, the rabbit is left in a carrier near the enclosure. We monitor via remote cameras to determine the monkey's level of curiosity and the dynamics of the initial interactions. The rabbit stays out of reach. If the monkey shows curiosity, the next step is to sit next to the enclosure with the rabbit on my lap. If the monkey comes over and begins to gently groom the rabbit, that's a good sign. If the rabbit doesn't mind being groomed or having the monkey playing with their ears, that's a great sign.

The next important step is to see how the monkey and rabbit will react during feeding times. Monkeys have a fairly rigid set of rules when eating together. The dominant monkey always eats first. If a subordinate eats out of turn, there are usually vocal remonstrations, possibly followed by slaps. Rabbits have no such rules, so we have to be certain how the monkey will react when the rabbit nudges them out of the way to get some vegetables. To get an idea of how a macaque might react to a rabbit at feeding time, I again sit next to the monkey's enclosure holding the rabbit, only this time I have the monkey's favorite treats in my pockets. I ignore the monkey and start hand-feeding the rabbit. As the monkey draws near with curiosity and begs for a treat, I hand them one and then one to the rabbit. This continues for a few minutes and if there are no adverse behaviors, then I'll put the food bowl down just outside the enclosure, so they both can reach it. If the rabbit eats while the monkey pilfers food from the bowl, that's a good omen and we will have a supervised introduction. If that goes well, then we gradually start leaving them together for longer and longer periods until they are full-time companions. We never push the process, preferring to let each monkey "tell" us when to move on to the next step. In Ivan and Maynard's case it took nearly two weeks to determine compatibility before the two were successfully and safely paired.

It was amazing how incredibly gentle Ivan was with his rabbit friend. He would give Maynard produce from his food bowl and then crouch on the ground next to him. With his face close to the ground and only inches from Maynard's face, Ivan would watch him chew his food. As Maynard finished each piece of food, Ivan would place another piece of carrot or broccoli in front of him and start the watching process all over again. This would go on for as long as Maynard ate. It was hysterical and always made me laugh.

Over the years, we've paired macaques and rabbits many times. Even though no two situations were alike, they were remarkably consistent in one way. Each macaque treated their companion rabbit very differently than they would another macaque. They allowed the rabbits to do things they would never tolerate from a monkey. Sometimes, while the macaques were on the ground eating from their food bowls, the rabbits would nibble food right from the macaques' hands or from their bowls. Sometimes the rabbits would even push their companions away from the bowl. Where a similar action would result in an all-out brawl between macaques, the rabbits were allowed to do this without a correction.

As much as Ivan loved his companion, I knew Maynard could not fully substitute for another macaque. Ivan needed a macaque companion to groom him and join him in climbing up the perches and hooting at the horses and deer in the meadow.

Then another research retiree arrived at the sanctuary. We knew Winslow would be a great candidate for pairing with Ivan, as his disposition, age and size were all perfect matches. With the arrival of Winslow, Maynard was retired to my granddaughter's house, to spend the remainder of his life hopping with the other rabbits she kept.

As it was when we introduced Maynard to Ivan, the introductions between Ivan and Winslow took quite some time. Yet, in the end we achieved a happy, successful pairing. Now, as I watch Ivan and Winslow chasing frogs and butterflies, dipping their toes in the swimming pool, or just lounging in the sun and grooming each other, I can't help but smile. Nothing in the world can warm my heart like watching a monkey enjoy life in a way I know they could never have imagined.

ABOVE *Ivan and Winslow enjoying a sunny day by the pool.*

Pearly Su

KEENLY observant, empathetic, able to learn new skills; these terms could be used to describe a top-notch corporate recruit. They also could be used to describe Pearly Su.

Like Ivan, Pearly Su came to OPR from a research facility. Unlike Ivan, who was born in the United States, she was born at a facility in China, dedicated to providing monkeys for research. When she was about 1 year old, she was shipped to the United States to be used for research. Then, when she was only 3 years old, still very much a juvenile, she was retired from research. While most research monkeys are used in multiple projects over many years—as Ivan was—Pearly Su was exposed to a single study.

Just like children, monkeys learn by observation and then trial and error. Pearly Su is no different, except she learns faster than any monkey I have ever known. Her abilities are nothing short of amazing and constantly keep us on our toes. Almost from the moment she arrived at OPR, Pearly Su took a keen interest in the locks on her enclosure. She watched me intently as I locked and unlocked the enclosure padlocks many times a day. Her interest was so strong that we began to joke about what would happen if Pearly Su got hold of the keys.

And then it happened. One day, as I went to unlock the padlock to her enclosure, I dropped the key. Before I could blink, Pearly Su's chubby little arm shot through chain link and grabbed the key off the floor. As I waited for my assistant to bring the extra key, Pearly Su sat in her hammock with the key making play faces and seemingly very pleased with now having her very own key. I planned on going into her enclosure and retrieving the key, but Pearly Su had other ideas. She went straight for the padlock with the key in one hand, determination in her expressive eyes and a free foot grabbing the padlock. To my amazement, she successfully inserted the key into the slot, and I heard the tell-tale click of the lock opening. She had opened it faster than I would have done so myself!

I was dumbstruck. No monkey at OPR had ever shown this level of learning. Fortunately, reality came back to me and I quickly moved to hold the enclosure door shut. It was good that I moved when I did. As soon as she opened the lock, Pearly Su was trying to open the door. I'm certain that had I not been there, she would have soon succeeded. After a few minutes, Pearly Su became tired of her game and, much to my surprise, handed me the key (just as the spare key arrived). Most monkeys will not share when they get a new object. Perhaps she realized that the key belonged to me and wanted to return it to its rightful owner. Still, the adventure with the lock and key gave me an idea.

Realizing that Pearly Su had such an interest in opening locks, I attached a padlock to the chain-link wall of her enclosure. I gave her the key and then watched. Quickly, Pearly Su took the key and went up to the lock. Holding the padlock in her foot, she intently studied it and the key. Within a few moments, she inserted the key into the padlock. After watching me open locks many times every day, it took her no time to figure out how to turn the key after it had been inserted. Less than a minute after I had given her the key, she had successfully used it to open the padlock. This soon became a game for us. I would attach a padlock to her enclosure, give her the key, and she would open it. The sounds that came from her upon opening a lock could only be described as "hoots of satisfaction."

But it was more than just padlocks that Pearly Su figured out. In her enclosure hung a mesh hammock, attached to the ceiling via carabiners on chains. The monkeys love these hammocks and spend many hours in them, watching the world from up high. At the same time, they make a terrible mess of them, so each day I take down the soiled hammocks and replace them with clean ones. As with everything else I did with Pearly Su, she watched my every move intently, sitting on my shoulder and carefully studying as I detached the dirty hammock and replaced it with the clean one. If my shoulder wasn't close enough to the action, I would soon have Pearly Su sitting on my head. Even if it was a little annoying, the intensity of her observation was such that I couldn't possibly disrupt her.

Then, one day after I had taken down the soiled hammock, but just before I hung the clean one, I got called out of her enclosure. Anticipating that I would quickly return, I left the clean hammock on the ground, ready to be hung when I got back.

An hour later, I returned and stopped, astonished, as Pearly Su looked down at me from the hanging hammock. Again, her "hoots of satisfaction" and play faces greeted me. She had figured out how to work the carabiners, hauled the hammock up, and managed to attach it to them. It wasn't perfect, but Pearly Su's ability to learn from observation had once again stunned me. To this day, no other monkey at OPR has ever figured out how to hang a hammock.

Pearly Su's skills aren't limited to inanimate objects. Her powers of observation have extended to the monkeys around her, particularly Ernie. It had been quite some time since we had tried to pair him with another monkey, as they were so unsettled by his seizure disorder. Even though they had witnessed these seizures many times, I could always tell when he was in the midst of one from the amount of distressful screaming coming from the nearby enclosures. Still, we were always hopeful that we could find a match for him.

Then, soon after Pearly Su arrived at OPR, Ernie had a seizure. As we ran to his enclosure to comfort him and make sure he didn't injure himself, all of the other monkeys were screaming in distress—except Pearly Su. Her concern was evident as she lip-smacked and cooed at Ernie. Once again, she had confounded our expectations. Right then, we decided to try pairing Pearly Su and Ernie.

It worked, and was an amazing match that lasted for over 12 years. Pearly Su was the comforting companion that Ernie had so desperately needed. She could sense when Ernie was about to have a seizure and would hug and hold onto him until he recovered, preventing him from falling or otherwise hurting himself. Between seizures, Pearly Su and Ernie lived a rich and event-filled life. Whether it was participating in the many activities available to them, such as hunting for bugs or swimming in the pool, they were always together. Or they would just sit in the hammock or on a ledge, grooming each other for hours. At the end of the day, they would snuggle close and sleep together for the night. They were always kind and affectionate to each other. In those 12 years, I never once witnessed a fight between them.

Then, after taking medication for his heart and seizure conditions for nearly his entire life, Ernie's health began to decline rapidly. Seemingly overnight, Ernie stopped eating, stopped playing, stopped hunting for bugs, stopped grooming. He could barely climb to the sleeping ledge to snuggle with Pearly Su. We called our veterinarian, who examined him, and gave us the sad news. Ernie was in liver failure. The very medications that had kept him alive for the past 13 years had gradually taken a toll on his liver. There was nothing we could do.

Pearly Su knew that her longtime friend was not well. She stayed with him constantly, even sleeping with him on the bed we had put on the floor of the enclosure. She groomed him so gently, cooing quietly as he lay on the bed. She would only leave him for the briefest moments to go to the bathroom or grab a quick bite to eat, then would rush back to his side.

Only a day after receiving the heartbreaking news about Ernie's liver, we received another sign that his health was failing. He had experienced a brief burst of energy and had climbed onto a ledge a few feet off the ground, when he suffered a cluster of grand mal seizures. Pearly Su grabbed onto him, comforting him as she always did, but couldn't hold on. He fell to the ground, scraping his head and legs.

We knew that it was time. It was an extremely distressing moment for everyone involved, especially Pearly Su. We had to pry Ernie from her arms so that he could be put to sleep. Everyone was crying, even our veterinarian. It was an overwhelmingly emotional moment that I shall never forget. And then it was over.

Afterwards, we showed Ernie's lifeless body to Pearly Su. Even then, her empathy for him remained. She tried to pull him back into the enclosure, cooing and lip smacking the entire time. As we left, her distress cries followed, forever haunting me.

Following Ernie's death, Pearly Su experienced a deep sadness. She refused to eat for several days; she would not play and would barely acknowledge my presence. After spending 12 years with Ernie, Pearly Su was mourning. The other monkeys must have sensed her sadness, as they would lip smack and coo whenever she was in their sight.

It wasn't until a year later, when a young macaque named Jack came into her life that Pearly Su healed. Jack had come to us after being discarded at an animal control agency by his previous owner, who had thrown him against a wall, breaking his tiny arm. His experiences had left him unsure of everything and distrustful of all people. Once again, Pearly Su saw a damaged soul and took on the role of the nurturing healer.

Within a few short weeks we were able to successfully pair Pearly Su and Jack. I began to see glimpses of playfulness from Pearly Su and it made me smile. Then, a couple of days after we paired them, they were sitting in the tunnel leading to their outdoor playground. As I neared them, Pearly Su stuck her feet out for me to play our game, "Where are Pearly's toes?" She had loved it when I would tickle her feet and play with her toes. She hadn't done this since Ernie's death. As I played with her toes, suddenly Jack's much smaller toes were beside Pearly Su's, waiting to be tickled. I cried a little at that moment, realizing that monkeys have to go through healing processes just like people. I knew that Pearly Su's heart had mended and Jack was on the way to regaining his trust in people. Now my heart, too, was healed.

LEFT *After Ernie died, Pearly Su would spend hours curled up in a ball, looking out into space. After 12 years together, she truly mourned the death of her friend.*

Epilogue

PROVIDING sanctuary for captive monkeys is completely different from rescuing wildlife. When injured raccoons or birds were brought to me, the goal was to nurse the animals to health as quickly as possible, and then release them back into the wild. That is not an option for these monkeys. A monkey who comes to OPR requires a lifetime of care, which can be over 30 years.

Whether the monkeys are retired from research or abandoned as pets, their history, personalities, physical conditions, and health needs are tremendously varied. The time, effort, and patience required to bring them back to a semblance of normalcy can be daunting. Yet, when one of these broken souls begins to become a monkey again, I feel such a sense of hope. Then, to watch them become a part of the larger monkey society at OPR provides me with an even greater sense of relief. It never ceases to amaze me how monkeys from such varied backgrounds and species can create their own social structure and rules. It is fascinating to watch and a privilege to be a part of their society.

The tales told in these chapters represent only a fraction of the individual monkeys' stories I have witnessed. And even for the personalities and circumstances described here, the long lives of the monkeys allow for many, many other stories to unfold, some deeply emotional, others incredibly uplifting, every one unique to that individual.

Summer

Three years have passed since Summer came to OPR. It takes constant work and vigilance, but she has shown remarkable improvement. She is no longer injuring herself, although some of her less destructive neurotic behaviors (slapping herself) will likely remain for the rest of her life.

After getting the worst of her self-injurious behaviors under control, we began to contemplate pairing her with another monkey. As social animals, it is vitally important that they have friends to play with, chatter to, and snuggle with at night. Unfortunately, with her behaviors, it has been quite difficult to identify the right monkey for Summer. Most monkeys get very upset when she begins to slap herself or display other odd behaviors. Then, there is the issue of her chronic vomiting and diarrhea.

Between Ernie's heart, liver, and neurologic conditions and Summer's neurologic and gastro-intestinal conditions, it often felt like I lived at my veterinarian's office. At first, we believed Summer's diarrhea and vomiting to be due to stress. She had come from such a dreadful environment. We were hopeful that the OPR environs and an appropriate diet would

ABOVE *After a long rehabilitation—both mental and physical—and a critical modification to her diet, Summer is now a confident and active monkey. She enjoys feeling the warm breeze against her fur and soaking up the healing sunshine.*

provide the cure. While it certainly helped, she continued to go through phases of vomiting and/or severe diarrhea, and started losing weight. So we went back to the veterinarian for more tests, more diagnoses, and more hopeful therapies. Summer had radiographs, ultrasounds, colonoscopies, and seemingly dozens more tests. She was always taking some medication. We were beginning to wonder if her health issues were like her behavioral issues, a product of her deprived upbringing. Then, after yet another exhaustive battery of tests, we finally determined her digestive system could not tolerate gluten.

With some trepidation, we removed gluten from Summer's diet and the results have been amazing. While on rare occasions she may still vomit or have a bout of diarrhea, the incidents have dropped to almost zero. Finally, Summer has started to gain her weight back.

With her health finally under control, we can now pay attention to the budding friendship between Summer and Tyler, who lives in the enclosure across from her. We have seen them holding hands through the safety mesh of the tunnels between their enclosures. Even though they are different species (Summer is a rhesus macaque and Tyler is a long-tailed macaque), they seem to get along very well and we are hopeful that they will be compatible and Summer will finally have that companion she so desperately needs.

Annie

It was a warm summer day and all of the macaques were outdoors exploring, playing in the pool, and chasing bugs, or just basking in the glorious sunbeams. Their happy grumbles and chitterings filled the air. Unfortunately, Annie was rarely allowed outside. She was so allergic to so many things. The previous spring, we had let her outside only to have the pollen in the air trigger a severe allergic reaction. After one frantic trip to the veterinarian, we decided it was best to allow her outside only when there was no wind and the pollen count was very low—a scenario that happened rarely.

There she was, sitting at her favorite window, watching the others play. I knew she wanted to join them, even if only for a little while. It was a bit early in the season, but it had been cloudy for so long, and beautiful summer days tend to be rare in the Northwest. I decided to let her go outside; although I would watch her closely to make sure she didn't start wheezing.

I brought Annie to the long wire tunnels that led from the indoor areas to the outdoor playgrounds, being careful to make sure that the doors to the playgrounds were locked. I couldn't let Annie into those areas, as the wood shavings would provoke an allergic reaction. It wasn't ideal, but I knew that she would still enjoy the feel of the air, the warmth of the sun, and the smells of the world around her.

It was a magical day for Annie, as she ran up and down the length of the tunnels (over 100 feet), chattering at the other monkeys, shaking the wire, and being the lively little monkey I knew.

It was with a little sadness that, as the sun started to set, I went to retrieve Annie from the tunnels to bring her inside for the night. I called her over to the access door. As she waited patiently while I removed the padlock, a wasp flew into the tunnel and circled her head. Without a moment's hesitation, Annie snared the wasp out of midair and popped it in her mouth.

I couldn't move fast enough and yet, it was as if I was moving in slow motion. I just knew that with so many allergies, a wasp sting was going to provoke a reaction. With panic in my head and heart, I quickly pulled her from the tunnel and took off running for the house. Annie was clinging tightly to my arms, sensing my fear and looking at me with concerned eyes. It seemed like hours, but it was only seconds. I was breathing so heavily, I could barely operate the telephone to call the veterinarian. Between gasps, I explained what had happened and that I would be leaving for his office immediately.

Then I looked down.

Annie was in trouble. Her breathing was getting raspy and forced. Her eyes were staring at me, wide with fear. I was screaming and crying, begging the veterinarian to help me, even as I knew there was nothing he could do. Her eyes slowly closed, as her little body began convulsing from the effects of the sting. I could barely see through all the tears. Annie, my companion and friend, went limp and died in my arms.

For a long time, I was wracked with grief and guilt. If I had only kept her indoors she would still be alive. Yet, I also knew that protecting her from every possible problem would have diminished her life. I consoled myself, knowing that her last day was so happy and enriched. I thought of the many years we had spent together, her incredible observational skills, and her devotion to me. I knew that I would think of her every time I drank coffee. Who would help me pour the creamer? I would think of her whenever I flossed my teeth. Who would pull my hair out to floss her own teeth? Annie will always be a part of me, kept safely inside my heart.

George

George is now nearly 7 years old. He is no longer the fragile infant sharing our bed. I share a very strong bond with him, but our relationship is changing.

He is a bit of a prima donna. Throughout OPR, we have hung many stainless steel mirrors in the enclosures. George loves to sit in front of a mirror and make faces at himself. There is no doubt that he knows the image in the mirror is his own, as he practices each expression in slow motion many times before showing it to another monkey. Sometimes, he'll turn his face one way, then the other, as he examines the expression from many different angles.

When we first moved George into the indoor habitats, it took many months for him to gain confidence. We were always there to help him— from the first day when he grabbed onto my arm so tightly he left bruises, until the first night that he slept in the habitat. George's mother would have been the one to teach him about monkey etiquette and social skills, but she almost certainly had been killed when George was brutally taken from his home in Thailand. Instead, we were his parents and teachers, doing the best we could to carefully and thoughtfully integrate George into our ever-growing troop of rescues and retirees.

ABOVE *In the main housing area, George loves to eat popcorn while he watches his favorite movies,* Elf *and* Uncle Buck.

Fortunately, George was a quick learner. He watched us interact with the other monkeys and quickly began to behave in a way that was socially acceptable for a growing male macaque.

Not long after beginning to live full-time with the other monkeys, George found a friend in Holly, a long-tailed macaque. On warmer days George and Holly spend time in the swimming pool, paddling around or chasing the frogs frequenting the area. Sometimes, we take a hose and put it in the pool enclosure. Just as children can spend hours playing with a hose, so will George and Holly. George has even learned how to put his fingers over the end of the hose and finds great delight in squirting any monkey or person who comes within range.

On cold or rainy days, George and Holly prefer to stay indoors and play with various toys or swing on the floating platforms or trapeze swings. They will sit next to each other for hours, hugging and grooming. George especially loves watching movies. He gets so excited when I begin to pop the corn, running around his enclosure and chattering loudly as the aroma fills the air.

As a young adult, George now commands respect. There was a time when I could enter George's habitat at any time and safely interact with him. Then, when he was about 6 years old, we were playing just as we always had, when he suddenly threatened me. I was a bit surprised, as he screamed and bounced at me. Not wanting to risk a confrontation, I quickly adopted a very passive stance, trying to redirect his intentions into a friendly game.

Most macaques at OPR view me as the alpha member of the group and George wanted to challenge that dominance. I understood that. To love George and his cousins is to respect them by embracing who they are, and by not asking them to be anything else. At the same time it was important that I not show him that I was frightened in any way. Macaques remember these interactions and, with the rest of the monkeys in the building looking on, I did not want to lose my standing with them. The tiny monkey, who had been torn from his mother and endured so much, was grown. He commanded and deserved respect. Soon, in fact, he would become the dominant monkey at OPR.

As I left his enclosure, all the other macaques were staring at us. I was concerned. Would they also begin to challenge me? How was this exchange going to affect the dynamics of the macaque barn? After nearly 20 years of providing sanctuary to monkeys, I still have so much to learn

about them. Perhaps that is why I enjoy the new and exciting challenges that mark every day at OPR.

It is hard to believe how far I've come in nearly 20 years. From that fateful day, when I first decided to bring Ernie into my home, to the decision to devote my efforts to providing sanctuary to abandoned and neglected pet monkeys, to accepting retired research monkeys; from the first home of OPR at our farm in Oregon, to our current 28-acre facility in southwestern Washington, I am constantly amazed.

There are many days where I look out over our property. I can look to the monkey buildings and hear the monkeys hooting and chattering to each other. I can look out over the fields, as deer wander through the lifting fog. In these rare moments of peace, I contemplate why I work so hard for these monkeys. Why would I continue to rescue pet monkeys from such tragic situations, knowing that they have years of recovery ahead of them? Why would I continue to bring in monkeys, retired from research facilities, who have never seen the sun? Why would I do this, knowing that all of these monkeys will require a lifetime of care?

The answer is simple. These monkeys need me. Until the time comes when monkeys are no longer kept as pets or used in research, there will be a need for sanctuaries like OPR to care for them and provide them with the life that they so greatly deserve. I have chosen a difficult path, but the rewards are tremendous. Watching George ascend to primacy of the monkeys at OPR or seeing Summer begin to form a friendship are such wonderful moments, they make all the effort worthwhile. Still, as I look out across the expanse of OPR, I dream of a day when there will no longer be a need for primate sanctuaries.

After reading these stories, you may ask yourself, "What can I do?"

OPR Coastal Primate Sanctuary is a USDA-licensed sanctuary, registered as a 501(c)(3) nonprofit. A lifetime of providing for the care and welfare of these beautiful souls is extremely expensive. We rely on donations to continue our mission to provide care in a humane and enriching environment for unwanted, orphaned, or crippled monkeys who originate from private owners, as well as those retired from a life in research. Please visit our website, www.oregonprimaterescue.com, to learn more about OPR, the work that we do, and what you can do to help.

Outside of supporting sanctuaries such as OPR, help is needed for legislative efforts to end the sale and possession of monkeys as pets across the country. Currently, in a majority of states, monkeys are allowed to be kept as pets with little or no regulatory oversight. For every Summer or Annie who find their way to OPR or other sanctuaries, there are so many others who continue to suffer in unsuitable conditions. Please contact your state legislators to voice your support for a ban on monkey ownership. For further information on this issue and how to get in touch with your legislators, please contact the Animal Welfare Institute at www.awionline.org.

We must speak for the monkeys, as they cannot speak for themselves.

POLLY SCHULTZ' fascination with monkeys dates back to her early childhood. Her love for and ability to understand and befriend all animals has always been a driving force in her life. Polly's unwavering dedication to helping animals in need led to the creation of OPR Coastal Primate Sanctuary (originally as Oregon Primate Rescue, in Dallas, Oregon) in 1998. Currently located in Longview, Washington, Polly spends her days caring for the monkeys and working hard to give them the best possible life. She lives onsite at OPR with her husband, Skip. Operating a sanctuary for special needs monkeys is labor intensive and requires constant supervision. Polly has not taken a vacation in nearly 20 years.

KENNETH LITWAK, DVM, PhD, is the laboratory animal advisor at the Animal Welfare Institute. He has authored or co-authored over 40 published scientific articles. This is his first book.